BOB'S *Wisconsin* DAIRY FRIENDS

Robert C. Bjorklund

Badger Books Inc.
Oregon, Wis.

Published by Badger Books Inc.
Printed in the U.S.A.
Cover photo by Arthur E. Peterson
All articles and photos originally published in the *Wisconsin State Journal*

ISBN 978-1-932542-33-2

Contact:
Badger Books Inc./Waubesa Press
P.O. Box 192
Oregon, WI 53575
Toll-free phone: (800) 928-2372
Fax: (608) 238-0495
Email: books@badgerbooks.com
Web site: www.badgerbooks.com

Contents

Introduction

During my husband's, Bob's, illness, I was searching for ways to entertain him. A friend suggested that I read some of his stories to him that he wrote in the Wisconsin State Journal. Bob enjoyed hearing them and remembering all the interesting people and places he had visited during his long career as farm editor.

As I read, I realized that taken together the stories represented a fascinating chronicle of Wisconsin's agricultural history. They covered a period of four decades that should be published in book form. Since it was impossible to include all of the stories, I chose to focus on those related to dairying, the most critical element of Wisconsin's agricultural economy.

Thus, "Bob's Wisconsin Dairy Farm Friends" came into being. I needed a great deal of help to make the book a reality. I am grateful to everyone who provided support and encouragement. In particular, I want to thank Marvin Balousek, the book designer and publisher; Dodie Remington and Donna Fisker, proofreaders; Lisa Miller, who shared her knowledge of the publishing business; Wisconsin State Journal staff members who helped Bob over the years edit and polish his stories; Ann Short, instructor of my Reminiscence Writing class, for her many helpful suggestions. I am also thankful to many other friends who provided moral support and encouragement.

Louise Archbold Bjorklund

Chapter 1

Milk Producers

Story Helps Farmers Start Anew

February 18, 1990

ARENA — Just seven and a half years ago, Robert and Pat Zinck of Ray Hollow Road were wondering if they would have enough cows to warm the barn and keep the water pipes from freezing.

Robert Zinck

They were just starting over in dairying and had just six cows and heifers in the big empty barn on their farm south of Arena.

But something happened that helped make their dream of a new start come true.

The *Wisconsin State Journal* carried a story describing the plight and struggles of the Zincks as "reorganized dairy farmers."

Jack Gorski, a UW-Madison professor of biochemistry who had a registered Holstein herd, read the story at the time he was looking for a different farm for his herd and for a new herd manager.

The Zincks had an excellent record as managers of registered Holsteins. When the farm recession of the early 1980s started to build and cut into the farm operations of many young farmers, Robert and Pat sold their herd of registered cattle to refinance their debt so they could stay in dairy agriculture.

About a week after the Zinck story appeared Gorski and Zinck talked about forming a joint operation. "Afterall, I had plenty of feed for the herd and a lot of time," Robert Zinck, 44, said.

Soon the Gorski herd not only was helping keep the

barn warm, but producing lots of milk. It was a beautiful seven-year partnership, Zinck said.

Today when you walk into the Zinck barn, all 50 cows are owned by the Zincks. By July they will have 60 head of their own and the barn will be full.

Zinck knew the quality of the original Gorski herd and bought about 26 from the professor. "I wanted to expand and be on my own," he said. He conceded he still misses Gorski's weekly visits to the farm to talk cows and bulls, the favorite topics of cattle breeders.

It has all worked out well for the Zinck family including Pat, who commutes to Madison each working day for her job with Dane County, and their daughters , Lori, an MATC student, and Darci, working at Lands' End, and son, Robert Zinck, III, a high school student.

What Zinck found is that "in troubles and trials and facing up to the strict economics of farming some good things can happen. Gorski coming to visit in the fall of 1982 was like manna from heaven."

"You have to have faith, keep plugging away and have a positive attitude," he added.

Zinck has been active in public and farm policy affairs of the Farm Bureau at the county and state levels, as well as with the Morning Glory Region of Associated Milk Producers Inc., his dairy cooperative.

In early March, Zinck will make his first trip to Washington with Farm Bureau members to meet with members of Congress.

His recommends that farmers who cannot personally participate in programs involving farm issues join a farm organization and have it represent them in dealing with elected state and federal representatives.

Robert and Pat grew up in the Boscobel area and began their farming career there on a rented farm. They then moved to the Montfort-Cobb area to work on a 50-50 basis in a dairy operation for the next five years.

They bought the 200 acre Arena farm in 1979.

As for the future of dairying, Zinck said he is more confident than ever.

"I'll tell you something, the farmers we've got now are real dairymen. Just the other day I noticed that of the top seven herds in Wisconsin three have an average production per cow of more than 1,000 pounds of butterfat, nearly double the state average," Zinck said.

"These are not 30 cow herds. These top producers have 120 cows in their herds," he said.

And the dairy situation is changing.

"We sold a lot of heifers out of the state and when you look at the Sunday paper each week, you see a lot more farmers are going out of business with an auction. I think we are going to be a lean milk production state for a few years," Zinck said.

Young Farmer Knows What His Future Holds

April 13, 1997

PLATTEVILLE — Mike Weigel, 19, is a young man with a mission.

He has made his objective very clear: "I want to own, operate and manage a beef-dairy farm."

Although the dairy industry has faced challenges in recent months because of weak milk prices, Weigel has not become discouraged.

In fact, when he thinks about his goals, he is more focused than ever.

Weigel said he is willing to face any challenge, knowing there are no shortcuts to dairy success.

Furthermore, he has the support of his parents, Steve

Weigel and Toni Johnson, — his uncles, who also have dairy farms — and the owners of two dairy farms where he works daily.

Support also comes from his instructors at the Southwest Wisconsin Technical College at Fennimore, where he studies dairy herd management.

You can tell where he is going because he says the best things he gets out of his classes are money management techniques. With that kind of knowledge, Weigel said, "I'll not buy something I can't afford."

His interest in dairying dates to his days as a student at Platteville High School, where, prior to his graduation in 1996, he worked to make the most of local opportunities.

His early efforts were rewarded with a $200 scholarship last year from the Grant County Veterinarians. The money helped offset costs of his dairy management studies.

"I worked on farms all my life and always wanted to be a farmer," he said.

Beyond the hours he spends in class, Weigel is up at 5:20 each morning to get to the Karl Klar farm, south of Platteville. There, he milks and feeds 40 cows, cleans milking equipment and drives his new, red truck 30 miles to Fennimore for his classes.

But the classwork doesn't conclude his day. After his afternoon classes, he drives to the Terry Peterson farm near Lancaster. Mike milks his share of the 165-cow herd there and completes other chores like cleaning the free stalls and the milking parlor.

In the summer, he helps his father complete field work.

He says Peterson's farm is an inspiration because of Peterson's cattle export business. Recently, the farm filled an order to ship 1,000 Holsteins to Mexico.

Weigel's farm experience has taught him how to operate all types of tractors, combines, skid loaders and other farm equipment. He also is an accomplished welder.

Asked if his full days of work on two farms and hours of study are too much for a young man, he hesitates, but says, "I'm working the way a farmer works — 5 a.m. to 9 or 10 p.m. seven days a week."

Long Shot Pays Off for Farm Family

March 26, 1995

CAZENOVIA — There's "a-million-to-one shot" farm enterprise on Route 2 that's working well, even though some of the neighbors admit they used to take bets on how long the new farm operation would last.

"No one gave me any encouragement, but I think my neighbors helped me a lot — just by saying I couldn't do it," the farmer recalls.

Starr Spencer has become a successful Richland County dairy farmer. But when she started on the Cazenovia area farm she was a city woman from Zion, Ill. She began coming here 20 years ago. Her farming venture started small, with $1,000 in savings.

Furthermore, her resolve to go into farming was strengthened because she had a confident — "We can do it together" — philosophy of what farm work and added responsibility could do for her two sons.

When the $1,000 didn't cover the $34,000 asking price for a 120-acre farm and some run-down buildings, she negotiated and got the farm on a land contract by offering $2,000 more than the asking price.

Today, Spencer has 600 acres on two farms and rented acreage, under the banner of Shiloh Farms. On one farm she milks 46 cows, the other is a beef farm.

Her two sons, Ben, 24, and Daniel, 18, farm with her and love it as much as their mother does.

Starr Spencer

Spencer and her daughter, Katie, 12, (who is crazy about horses and rodeo) live on one of the farms. Another daughter, Jennifer, 23, trained as a veterinary technician (at MATC in Madison), is working in a vet's clinic in the Grand Cayman islands.

Just how did Spencer find a farm in Richland County?

Her parents took the family on a vacation in Wisconsin Dells each year and they would travel the area for new experiences. Spencer was attracted to eastern Richland county.

After buying the farm, she and her children traveled from Zion to the farm every weekend. She wanted to see if they were interested in farm life. It worked. They had early success with a corn crop, which enabled them to buy their first tractor.

The first real expansion was to take on a Red Angus beef herd to help out a brother-in-law who had the herd and needed a new farm to hold it. The Spencers took over and were paid for their service and feed with calves from the herd.

Today Ben continues with a 60 cow beef herd of his own. He has built a Red Angus reputation and has repeat breeding stock buyers from southwest and northern Wisconsin.

Starr and her family moved to the farm in 1984. By that time her sons were convinced about being farmers and told their mother they would stay even if they had to sleep in a tent.

Spencer is a Farm Bureau member to improve her knowledge of farm programs.

She says all people should be more agriculturally aware and city people should know about farmers and what decisions they have to make.

When Spencer's relatives learned she was farming and had decisions to make on ballot initiatives issues involving pesticides in California, "they called me to ask me how to vote and what decision was good or bad for farmers."

Jerseys are his Bread and Butter

Oct. 8, 1989

Janesville area family farmer and Jersey breeder William Barlass is at World Dairy Expo with his 1989 show string of Jerseys because expo is his best contact with foreign buyers.

This contact, he said, has put Barlass Farms Jerseys on farms in six South and Central American countries, including Colombia, Peru, Brazil, El Salvador, Equador and Costa Rica.

"Expo has become our outlet to the world," he said. The reason: Demand for Jersey cattle in the past 10 years, he said, has grown steadily because of the breed's ability to adapt to the heat and insects common to those countries.

In addition, his buyers report that Jerseys are efficient converters of forages to high quality milk in South America.

What's more, he has a 100-cow registered Jersey herd at the home farm east of Janesville for the breeding stock needed for Expo visitors, both foreign and domestic.

Barlass, 37, a UW-Madison dairy science graduate, grew up with the Jersey touch during 16 years of work with Gil-Bar farms. This recognized name in cattle breeding was started 62 years ago by his grandfather, Alfred Barlass, and his brother, Stewart.

After this start with an established name in cattle breeding, Barlass bought a farm in 1968 and the following year started his own operation with six cows.

His foundation herd traced back to his first 4-H heifer, named Hetty, a registered Jersey and his six cows, plus another 16 cows owned by his father, Marvin.

A dramatic herd expansion came in 1977 when the barn was enlarged to handle 100 cows.

All of the heifers from the herd are raised for replacements or sale and up to 30 head of calves, heifers and young cows are sold from the herd each year. In big years, Barlass has sold 50 from his farm to foreign and domestic buyers.

In the best years, some Barlass cattle sold in the $10,000 to $15,000 range.

Barlass Farms is a family enterprise with Bill, his wife,

William Barlass and one of his prize Jerseys.

Marian, and their three children, Brian, 13, Kristin, 9, and Brett, 5. Marian Barlass served as a home economics agent with Dane County Extension. She also is a UW graduate and had the Jersey touch since she came from a Jersey farm at Frederic in Polk County.

For 10 years William and Marian were editors of the *Wisconsin Jersey Booster*, the state breed publication. As for the dairy farm future, Barlass said he is optimistic. "I've been concerned about the number of cows cut from herds and the number of family dairy farms that have gone out of business," he said.

"But we see a very strong demand for the milk and manufactured dairy products like cheese, yogurt and ice cream," he added. He especially noted that government dairy reserves are at the lowest point they have been in years, with butter as the only surplus.

"From the statistics we are seeing," he said, "it looks like there will be a shortage of fluid milk," Barlass said.

No Pigs in a Poke at this Farm

October 14, 1990

OREGON — When you get south of here on Glenway Road, where Locust Grove Road intersects, you are at a place once known as Dane Corners.

Not many people use the name today, but one rugged Dane, Lyle Nelson, 74, whose farm is right at Dane Corners, recalls the cluster of Danish families that farmed in the area.

"This community of Danes had names like Jensen and Fredrickson — and Nelson. It was a settlement like the Norwegians had in Stoughton and Mount Horeb or the Swiss at New Glarus.

"But many of the Danish immigrants, including my relatives moved on, first to Minnesota and then to California.

"My grandfather and grandmother came here from the Copenhagen area of Denmark and many relatives followed them." The Nelson farm, which specialized in dairy cattle and hogs, started its registered Holstein herd in 1918 and continued until 1966 when Lyle's father died and the herd was sold.

His father served for years as town treasurer and as president of the Oregon Mutual Fire Insurance. Lyle still has the certificate showing his father as the valedictorian of his class at Oregon High School.

His mother also was active in the area and was an officer of the Grange.

"When I think about it, I got a good start in life because of my parents. For a time I hired out to farmers and worked for 50 cents a day. I had other opportunities to work on farms, but I figured dad needed me here.

"I was so busy I didn't get married. It is something you keep putting off. And I didn't have the money to do so. It was the dirty '30s — the Depression," he continued.

"That's when hogs sold for $2.55 per 100 pounds and I sold 80 purebred hogs at that price. We raised tobacco and got 2 cents a pound for the crop. Today it's $1.25 a pound," he said.

But Nelson showed his abilities in Holstein cattle and, surprisingly, in poultry. In fact as a 4-H member he was a state winner in poultry judging and in the national competition in Dallas.

Later he was elected president of the Dane County Holstein Breeders and handled all the details of registered cattle sales for the late Bert Pfister, a Mount Horeb auctioneer.

He also was president of the Dane County Dairy Herd Improvement Association and secretary-treasurer of the Dane County Pork Producers.

For 40 years Nelson made the selection of cattle and calves for the Dane County Holstein Breeders Opportunity Sale to help get young people started in registered cattle.

He also was a recognized cattle judge and worked with 4-H youth on showing and fitting cattle. He was a 4-H club leader in his neighborhood.

Nelson said that when he finishes his current term as vice-president of the Wisconsin Duroc Association that may be the last office he will hold.

"I think I did my turn," he said.

He also is active in the Optimist Club and the Oregon FFA Alumni.

When a visitor comes to his farm, Nelson is likely to ask: "Say, would you like to see some really good hogs?"

He'll walk to a pen of his favorite young pigs, which scramble to the far end of the pen as you enter.

But then Nelson uses his hog magic and says in his soft voice, "come on guys, come on. Let's see you."

"Just look at that gilt isn't she perfect," he says with

pride. The sleek and growing July and August pigs will move toward Nelson.

When they are as close as they can be they stretch and arch their necks to look up at Nelson just as if they were reporting in for inspection.

There are more than 20 gilts in the pen and 14 are sold already.

Then Nelson moves to the pen of his favorite sow, one that gave birth to 16 pigs in her last three litters and raised all but one or two.

"She will never lay on a pig even though she's a heavyweight," he said. And then he shows with his hands how slowly she will settle down so her pigs can nurse.

The results of pig breeding have been rewarding and recently he had the champion Duroc market hog at the Belleville Test Station. At the State Fair he has had a string of honors.

He supports young exhibitors in 4-H and FFA by selling his barrows for their swine programs and charging only a minimum price for quality swine entries.

But he wants to see those youngsters have a chance to win. One of his Duroc barrows shown by a young exhibitor in Racine County sold for $1,300 at the fair this year and another was the champion Duroc barrow at the Rock County 4-H Fair.

He has been such a hit with young people "they tell me I can't quit. He had 14 young people buying barrows last year.

Nelson farms alone and has no hired help during the year. But he still has relaxing moments turkey and deer hunting and getting out in his new boat to do some fishing.

Farmers' Hard Work Rubs Off on Kids

October 23, 1994

RIDGEWAY — Someday, someone is going to have to recognize Art and Helen Kirch of Rumpus Hollow Farm for the part they've played in getting other family dairy farms off the ground.

Art, 56, a born dairy enthusiast, says if you spend enough time working with young people, they can establish successful dairy farms even in today's competitive climate.

He knows his system works.

Art and Helen have helped create four dairy farms — run by the families of their four children. The Barneveld area farms are operated by daughter Barbara; her husband Ted and their family of three; son Tom, his wife Rhonda and their two children; son Nick and his wife Maria; and Art Jr.

A fourth son, Bob, does construction work in the Milton-Edgerton area.

Art's goal of getting his children started in the dairy business took more than a positive attitude, he says. It also took a lot of hard work.

Art, as a businessman, operates with this simple philosophy: "I demand that things on the farm be done right."

Also, "you've got to like farming and you've got to make money at it or you don't stay with it too long."

Apart from a good work ethic, Art says he worked hard to instill a sense of financial discipline among his children. That discipline involves putting away the money to pay the bills and the property taxes.

And Helen adds that the little things Art requires, like sweeping up spilled feed, caring for the cattle, feeding them and sweeping the mangers also helps ensure success.

"It all makes a difference," she says.

To help young farmers get off to a good start, Art follows a well-defined strategy:

His son or daughter (or a hired man in one case) start off working one of Art's farms with his herd of cattle.

As their progress becomes evident, Art begins offering a percentage of the farm's income.

That step is taken so they can move into the third step, which means buying half of Art's dairy herd on the learning farm and half of the farm machinery.

Finally, the person in training reaches the point where he or she is able to rent and operate the farm.

Ultimate success is reached when the trainee buys the farm and begins mapping out his or her own plan for the future.

At that point, Art faces his most difficult period. "I have to give up control of the farm," he says.

Art started his career near Arena in Iowa County at a farm on Highway K, south of Highway 14. He met Helen

Art and Helen Kirch

Marty, at a dance and they were married 34 years ago.

From Arena, Art and Helen moved to Vondron Road on Madison's far East Side — where the new World Dairy Center and the new State Department of Agriculture building are being developed.

The couple rented two farms there for nine years and had the last dairy operation on Vondron Road.

Art says he thought of buying the farms on Vondron Road. The land was going for $800 an acre then. It recently sold for $25,000 an acre.

Instead, Art chose to move to Iowa County's town of Brigham, near Barneveld, where he and Helen bought two farms and 800 acres.

They soon learned that sweet success in farming can change dramatically when the tornado that hit Barneveld on June 8, 1984, also hit their farm, destroying their two barns, five silos, and damaging their house.

Fortunately, they were able to rebuild and repair. Art has one wish today. He would like to work with a firm or an agency that lends money to new farmers.

Art says he believes he has something to contribute to those new farmers. He would like to pay periodic visits to such farms and work with the new farmers to build and improve their operations.

"Farming is hard work, but there is money to be made in the long pull," he says.

Winfield Farmer Enjoyed Four Careers

November 11, 1990

REEDSBURG — The four career interests of town of Winfield farmer Robert Whitty, 70, tell his story.

Whitty, first of all, is a farmer with his wife, Virginia

— a team that excells in leadership, innovation and management skills in both dairy and dairy beef operations.

Then there were his studies and 1942 graduation in agriculture at the University of Wisconsin, followed by five years service as a single engine P-47 fighter pilot in the Army Air Force in the Pacific in World War II. He was commander of the 456th Squadron.

Even Virginia had a wild flight in a two-seater P-40 when he was a flight instructor.

When he was discharged as a captain in 1946 he started another career as a vocational agriculture instructor at Adams-Friendship High School. He not only taught, but was a factor in starting the Central Wisconsin Junior Livestock Show and then the Central Wisconsin Beef Producers sale program.

After 10 years as a teacher, it was time to return to farming and he purchased the farms of his father and uncle, who worked 280-acres as one farm. Included in the acreage was the original 80 acres of the Whitty Farm on Highway V, four miles north of Reedsburg.

This farm has special meaning to the family because it was settled by Whitty's great-great grandfather, Patrick, in 1843, who came from Ireland.

Community service in the Reedsburg area followed his return to the farm when he served 24 years as a member of the Reedsburg Board of Education, including 17 years as president.

Whitty makes it clear that all four careers, farmer, fighter pilot, teacher and school board member are woven into the lives of both Virginia and himself as a result of his education at the University of Wisconsin.

"I really feel indebted to the university for all it has done. The university was important to me, wherever we traveled," he said.

Today, Robert and Virginia are proud that all six of their children are graduates of UW-Madison.

Virginia adds that their oldest grandson is a student in the UW-Madison School of Business.

The university also has recognized Whitty, who was cited this year as a recipient of the honorary recognition award of the College of Agricultural and Life Sciences.

Robert Whitty

He also holds the Wisconsin Agricultural and Life Science Alumni Association's distinguished service award.

As an Alumni Association director he was on the committee that developed the highly successful Football Fire-Up for agriculture alumni each fall.

Whitty recalled numerous times he received the counsel and encouragement of Agriculture Associate Dean Vincent Kivlin, and later, Associate Dean George Sledge.

While a teacher, Whitty would spot students with promise and Kivlin was able to get them started at the college. Kivlin later wrote Whitty that "if you have any more of those jewels in the rough up there, bring them down."

When Witty was at the UW he worked in the Hill Farms university dairy operation where the first milking parlors were being researched.

So it followed that he installed Sauk County's first milking parlor and free-stall housing system for his dairy herd.

Whitty also was a promoter of Holstein dairy beef feeding operations because they fit in so well with a basic dairy farm and feeding program. The Holstein steer will balance and diversify farm operations.

Today the Whitty farm covers 500 acres well protected by excellent soil and water conservation plans, and Whitty's son, Mike, is taking over the operation.

There is a lot of harmony on the farm today because

Mike's wife, Robin, is director of a UW star attraction, the University Singers.

And Robert and Virginia proudly recall that Reedsburg has sent more singers to the University Singers than any other high school. Robin and Mike were singers for three years. Robin later was music director at Reedsburg.

"My son and I have come to the conclusion that when you look back at the 100 years our family has been on this farm, they certainly have had ups and downs along the way. We aren't going to let the downs discourage us too much, but we are going to enjoy the ups, too.

"I think, maybe, farming has to be looked at as a long haul operation and is something that will have its ups and downs," he said.

Two-Story Dairy Farmers Get National Recognition

December 16, 1990

YUBA — At the top of the hill on Highway 56, just six miles southeast of Viola, there is a family dairy farm with a future.

It is the headquarters of Junlyn Farms of three Kepler families — Evelyn and Carroll Jr., who are the senior Keplers and own the land, and their sons, Kevin, 36, and Ron, 39, both with young families, who are operators of the 960-acre enterprise.

The Keplers' world centers on the milking herd of 120 registered Holsteins.

This is a new farm.

The entire farmstead, including three houses, was built on land purchased in the early 1970s. Only one building, a small barn, was kept from the previous farm and was remodeled as a calf barn.

The Keplers are a family with goals, experience, wisdom and love, the qualities all expressed by the senior Keplers in a lifetime in dairy agriculture.

They wanted this for their family. Kevin and Ron both wanted to farm and their parents gave them a chance.

Today, it is not only the farm that has attracted national and international attention, but it is Carroll Kepler Jr., and Evelyn Getter Kepler, both 65, who have become recognized personalities in U.S agriculture.

The Richland County couple is one of the 16 select farm families chosen by the editors of the 1990 Yearbook of Agriculture, "Americans in Agriculture: Portraits in Diversity."

Their farm story is published in the book recently released by Secretary of Agriculture Clayton Yeutter. The secretary called people the "richest resource of American agriculture."

The 200-page book was published by the U.S. Department of Agriculture as one of its annual series and is dedicated to help people understand the character, philosophy, struggles and accomplishments of farmers dedicated to their families and the land.

The Kepler story, written by UW-Madison Agricultural Economist William Saupe, is one of the profiles in the book to give readers the real meaning of what it is like to be a farmer, specifically a dairy farmer in Wisconsin.

The portrait collection includes men and women in the business of agriculture, agricultural scientists, those in agricultural education, government workers in agriculture and volunteers supporting agriculture at local, state, and national levels.

Readers will get to know the Keplers, who were married in 1947 when Carroll was working on his father's farm near Sabin and his herd had 25 cows in the milk line.

In 1952, they bought the Sabin farm and farmed there until they moved to the new place on the hill.

Saupe tells what it is like to farm a "two-story farm" with the bottom lands and the ridge tops tied together with field roads.

This part of Wisconsin is difficult to farm successfully, but Carroll and Evelyn and their family have succeeded, Saupe wrote.

Kevin and Ron, the new generation are continuing the Kepler story by telling how their father started buying the land for the new farm so those children of the family who wanted to farm would have the right place to do it.

"Dad wanted good land and that's what he has got," Kevin said.

The farm name, Junlyn, is the prefix of their registered Holstein breed name and formed by the combinaton of "Jun" for Carroll Junior and "lyn' for Evelyn.

Today those Junlyn Holsteins are in herds from Saudi Arabia to Brazil and China as well as dairy farms throughout the U.S.

Their herd and farm is a destination of visitors from dairy areas throughout the world and the nation. Kevin and Ron reflect the pride they have in their Holsteins because they developed 29 cows that have been classified as excellent.

The Keplers got into registered Holstein breeding when their daughter, Connie, now Mrs. John Turgasen of nearby Sabin, had a purebred heifer as her dairy project in the Sabin Livewires 4-H club.

Kepler believes that in this area of Richland County the way to get ahead in farming is to develop a good dairy herd.

That's what the Kepler sons developed and their cows are producing a herd average of 20,000 pounds of milk a year.

Kepler's life as a farmer included work as a farm management aide in the county for nine years. His work was with small, beginning and limited-resource farmers.

He is praised for his work by former UW-Extension farm management agent Maynard Nelson because Kepler's advice and encouragement helped people having problems stick it out in the troubled financial times of the 1980s.

His sons, Kevin and Ron, along with their wives, Joan and Julie, and their children also have caught the Kepler spirit. The farm is run as a family corporation with Kevin as president, Joan as vice president and Ron as secretary-treasuer.

"The farm operation is getting better all the time," Kevin said.

Kevin and Ron agree that the goal is to make money, but the last 10 years have been a struggle.

But the record shows they upgraded their equipment and paid down their debt and that, to, is "the way it has to be."

For Farming's New Breed, Look to Wolf Haven West

April 1, 1990

MAZOMANIE — You get a good measure of Simeon and Heidi Wolf's devotion to farming from the attractiveness of their farmstead, their emphasis on soil conservation and water quality and their skills in improving the production of their 30-cow Holstein herd.

They say they just want to make things better on their 164-acre farm, Wolf Haven West, on Hudson Road, just west of Mazomanie.

"Having 90 cows would be nice, but we want a herd of the size we can handle together. We want excellent cows and not more cows," Heidi said.

The Wolfs have been recognized as winners in three categories of the MATC Farm Business Production and

Management Program for 135 young farmers in Dane and Green counties.

Simeon and Heidi Wolf

What the Wolfs demonstrated with their farmstead, dairy herd improvement and soil conservation was their willingness to do the work and carry out goals set during the late 1980s, the most trying economic times for agriculture.

How trying? After Sim and Heidi were married in 1986 and decided to operate their farm separate from his family and brother, they were told, "You'll never make it on your own." They've proven otherwise.

Last year, milk and butterfat production records for the Wolf herd showed the greatest increase of any MATC program member.

Production per cow was up 4,205 pounds of milk and 131 pounds of fat. Average yearly production per cow is almost 18,000 pounds of milk.

The Wolfs' spectacular result came with forage analysis and ration balancing. The herd is fed seven times a day. Their artificial insemination breeding program is paying off, too.

There are two special animals in their herd. One is "Wedding Belle," a registered Holstein that Heidi's parents gave as a wedding present. The other is "Jingle Belle," the heifer's first calf, which came on Chiristmas Day. Wedding Belle has another calf due at the end of April.

Probably the biggest factor in Wolf Haven West's early success was that Sim was able to buy used farm equipment relatively cheaply because people were selling equipment to keep their farms operating.

Sim said he was able to buy a full line of equipment for $10,000 — about one-fourth of used equipment prices today. He also has good neighbors, who let him use their equipment for special needs.

The Wolfs' soil conservation is intended to keep Black Earth Creek, only 60 feet from their barn, as clean as possible.

The Soil Conservation Service designed a special filter system to keep manure from entering the waterway. It includes a 40-by-70-foot settling basin of bluegrass where the water can soak in without going into the creek. The gently sloping basin is mowed regularly during the summer.

Sim, 34, is one of 13 children in the Frank and Eva Wolf family. He attended UW-Stout, completed the Farm and Industry Short Course at UW-Madison and has strong mechanical and automotive skills.

Heidi, a 1987 graduate of UW-La Crosse in community health education, works in surgical admissions at Dean Medical Center in Madison.

During haying season, Dean allows her to take vacation days to help her husband.

"I spend my vactions in the hayfields, but you know how it is: you have to make hay when the sun shines," she said, laughing.

Heidi is an accomplished horsewoman from the Fremont area in Waupaca County and hopes to soon be boarding horses in a new, seven-stall horse barn. The Wolfs' farm is linked to an extensive network of riding trails.

Sim and Heidi are up about 5:30 a.m. each day and milk the cows together. Heidi feeds her horses and then is on her way to Madison for work. She gets home in time to be with Sim for the evening milking.

Heidi also is a floral enthusiast. Each spring and summer, Wolf Haven West bursts with color, including window boxes on each of the nine windows of the barn, hanging baskets and flowers in every corner.

The flowers, improved landscaping, well-painted and remodeled buildings, and an uncluttered farm yard, she said, show people "we are serious about making the farmstead one in which they are both happy and proud to come home to."

"Even the neighbors stopped to tell us how good it looks," Sim added.

Woman Farms in Male-Dominated Field

July 15, 1990

MINERAL POINT — If you didn't know it before, Lafayette County is very different than the rest of the state.

Kristie Popp, who farms with her husband, Dennis, on their 293-acre dairy farm, says "down here we are independent people and we like that feeling."

Popp, who is also an instructional aide in special education in the Platteville School District, says Lafayette County's independence is evidenced by farm policy developments, especially by getting women involved in policy decisions.

The way she describes that independence with an enthusiasm that Lafayette floods can't wash away or two years of drought can't burn out.

"Am I a radical person? I am. I know. But I am not a women's libber or anything like that," she says with a chuckle.

Popp is also vice president of the Lafayette County Farm Bureau, which makes her head of the policy development committee.

"There are so many women who separate themselves

from the policy field believing it is the man's thing to do, while the woman's thing is something like craft demonstrations or pouring the coffee," she said.

She broke that mold by gaining organizational experience while serving as county secretary-treasurer and chairman of the women's committee.

"I feel we go into Farm Bureau as a family and have to be actively involved as a family. We have goals to make a family organization work and that's where women become involved," she said.

Popp is the daughter of Robert and Betty Swart who have a dairy and hog farm north of Platteville.

She holds a bachelor's degree in social science from UW-Platteville. The Popps have a son, Nathan, 12.

The Lafayette independence Popp describes is most noticable when she talks of government farm programs.

Currently the world is trying to find a way for major industrial countries to reduce farm subsidies and Popp hopes this can happen.

She supports this policy change in the United States because she is confident farmers would handle commodities without government programs to acquire surpluses and then use them to lower farm prices.

"Farmers have shown more responsibility in farm product marketing," she said. "They also have their own marketing agencies, like the Wisconsin Milk Marketing Board, that can help with dairy promotion and fund research for new products."

"The key to success for American agriculture to prosper in a world with reduced subsidies is its efficient production. We want this efficiency to continue so our products can bring prices higher than our costs," she explained.

Popp knows what it is like to keep from using government programs since the only special program used on the Popp farm was the drought emergency act in 1988 and 1989.

There is a dairy program, but the Popp's 43 cows like all others have been producing for a free market for the past year and price supports have not been required except for butter, the only dairy surplus. Dairy programs do not involve the individual farmer.

The Popps bought their farm under terms of the Farmers Home Administration loan program.

"We've been farming here since Jan. 3, 1975, and will have it paid for in 4 1/2 years," she said.

Both she and Dennis would like to see Nathan take over the farm one day, but "it all depends on whether or not he wants it."

This decision will be worked out by the family, but Popp said Nathan should get the education first at a tech school or a college.

As for the future, Popp said the key policy issues farmers face are tax reform and education.

At the rural community level, she added, the issues are availability of doctors and clinics for health care and rising health care costs.

Auction Brings End to Fitchburg Farming Tradition

February 2, 1992

A proud and significant part of agricultural history in Fitchburg and Madison will start fading on Feb. 10.

That's the sale date for Lyman McKee's Farms, a farming operation south of Madison for 160 years.

McKee's grandfather, Dennis McKee, started farming in 1860 on what now is McKee Road.

Three McKee brothers left Ireland during the potato famine, and Dennis was one of the early group of Irish immigrants to settle in Fitchburg.

Today, Lyman McKee, 81, is a resident at St. Mary's Care Center, 1347 Fish Hatchery Road, 3.5 miles north of the McKee farm.

He spends his days quietly in his room, reading or using a typewriter to send a message or two to his friends. There are visitors who enjoy recalling some experiences they enjoyed with McKee.

His voice is strong, and he sounds like the McKee of old. The visitors admire McKee for his lifetime of leadership in the dairy industry — from the farm to the dairy plant and to the marketplace.

"I always had a feeling in the back of my head that if we could advertise and promote milk we could sell the world dry of milk," McKee said, reflecting his many years as president of the American Dairy Association of Wisconsin and president and general manager of Madison Milk Producers, a cooperative.

His dream of "selling the world dry of milk" never materialized, but there were some years — such as 1989 — when the world had to deal with an extremely short supply of milk and its products.

Even though his cluster of three Fitchburg farms was the source of thousands of tons of quality Grade A milk, the real contribution to dairy agriculture and national and world leadership was McKee himself.

Naturally, as the sale date approaches, a touch of sadness comes up in McKee's conversations.

"I really don't want to see the farms end this way, with me here and not on the farm. But I am getting wonderful care and think how lucky I am to feel this way, when others have greater problems. . . . I'd rather be with the cows," he said as his voice trailed off and memories took over.

But he is a realist: "I only hope we have a good day for the sale."

McKee is pretty well limited to his room at the care

Lyman McKee

center. But he thinks of his farms daily.

In fact, he is talking on the phone with people in the grain trade about leasing dryers and a 125,000-bushel storage facility.

Actually, the auction of machinery is the first step of a phased-out closing of the dairy operation. Jerome Ziegler, who has been with McKee for 36 years, will keep the milking herd in production until spring. Ziegler said there is plenty of feed for the milking herd and probably it will be the best move to disperse the herd in spring.

When McKee thinks of his dairy herd, he recalls that

he and his brother Bill never missed a single day in more than 30 years of working together.

Ziegler is going to stay on with the McKee enterprise, which is developing the farmland into residential and institutional properties.

McKee never enrolled at the University of Wisconsin as he planned because of his father's arthritis. The farm was McKee's education, and he was an excellent writer and speaker.

He also tells how proud he was to be appointed by Agriculture Secretary Orville Freeman to an advisory committee, serving with one Western cattleman, self-educated like McKee, and nine professors, all holding doctor of philosophy degrees.

Freeman would listen to the experts and then turn to Lyman and ask, "and what does Dr. McKee think of that proposal?"

McKee's grandfather's decision to settle in Fitchburg was a fantastic decision when you see how the McKee lands became so valuable with urban development. McKee still remarks how significant it was that "my grandfather chose this land right on McKee Road."

But McKee has a story to balance that, telling of how he took a small part to a farm equipment firm for repair and within minutes it was done.

"How much do I owe you?"

McKee asked. "Aw, 25 cents," the repairman replied.

McKee searched his pockets and purse. Nothing was there.

"What, you're farming and you haven't got 25 cents?" the repairman exclaimed.

"I guess that's right," McKee answered.

Farm Changes Direction After 148-Year History

Nov. 27, 1994

COTTAGE GROVE — Last week as relatives gathered for Thanksgiving on the Hammond Farm, 3859 Vilas Road, they went directly to the family manor, built in 1846 — two years before Wisconsin became a state.

It is an exceptional place to celebrate Thanksgiving and even though this special day was designated later, the Hammond family has been giving thanks for its food production for each of its 148 years.

The home today is the main house on the farm and the residence of Harry and Millie Hammond.

He's proud of the home's solid contruction his great-grandfather did more than 148 years ago using 8-inch by 8-inch hand hewn oak timbers from the farm woods for basic construction.

"If I ever got a tornado warning and I was in bed, I'd stay right here. I know I would be safe in this house," he said.

Harry, 70, a family historian, is the fourth generation of Hammond*s* to live in the town of Cottage Grove, a few miles south of the village.

A smaller quonset home on the farm has the fifth and sixth generations — Harry and Millie's son Joel, his wife and their children.

Joel often wondered if he would have room to raise his family of four in the small quonset.

But Millie said he felt better about it when she reminded him she and Harry raised seven children in the quonset.

The original Hammond farm was a 240 acre property granted to Alexander Hammond by a U.S. government patent in 1845. The Hammond family came from England, where Thomas Hammond was born in 1603 and settled in Massachusetts in 1635.

The farm has changed little since 1933, when 9-year-old Harry started farming regularly with his father. So Harry has been farming for 61 years.

Today he has a partnership with Joel and still does much of the planting and harvesting.

But, Harry said, the farm likely will cease its dairy operations in another two years.

Joel is milking only 24 Holstein cows (less than half the average herd size in Wisconsin) and Harry says, "there are so many changes in farming today that they are pushing small farmers out."

"I did better moneywise when milk was $3 to $4 a hundred pounds than I'm doing now with the all milk price $7 to $8 higher," he said.

Harry added that farm taxes and operating expenses are higher and the only reason they are able to continue farming is that they are using the same farm equipment Harry has used for 20 to 30 years.

"I am sad to know the dairy operation will end." he added. However, Joel said he plans to keep the farm going by raising Holstein steers.

Harry still raises two-plus acres of tobacco, but has no idea what the price will be compared to last year's $1.45 a pound. He sold it for as low as 16 cents a pound and his father got only 2 cents a pound in the tough days.

Recently, Harry was honored with a clock trophy and citation at the district meeting of Associated Milk Producers Inc. for his 50 years as a dairy cooperative member, first with Madison Milk Producers Cooperative, and the last 25 years with AMPI.

Retirement Ends Family Tradition

January 7, 1990

BARNEVELD — For Harland and JoAnn Rue, 1990 is a big year for changes in their lifestyle. For 41 years they have worked together on a 200-acre dairy farm in the beautiful hill country, kept that way by their soil conservation practices that trace back to those laid out by the CCC in the 1930s. The farm is south of Barneveld in Iowa County's town of Brigham.

The Rues have decided that this is the time to retire from the farm, which has been in JoAnn's family and farmed continuously since 1851 by her parents and grandparents.

The Rue (pronouced *rewee*) farm is steeped in family tradition because JoAnn's grandparents, who first farmed the area, aquired enough land to give each of their four daughters 200 acres, including JoAnn's mother, Esther. The Rue farm was the only one that continued in the family.

Now that will change, too.

The Rues are scheduled to sell their dairy herd and farm equipment the week of April 16. They plan to continue living on the farm, holding the land in the Conservation Reserve Program.

The Rue's four children all chose different careers. Three have received university degrees and their youngest daughter, Pam, soon will complete her studies at UW-Whitewater.

Harland and JoAnn started farming as partners when they were married in 1948. They began with a small herd of cows and delivered their milk to the nearby Prairie Grove

Cooperative cheese factory.

One thing Harland treasures is that he is still using all but one of the four tractors he acquired since he started farming.

The Rues soon got involved in many aspects of farm and community organization activity.

When the Rues started sending milk to the cheese factory they became active in dairy cooperatives, including many years with Harland as a corporate director of Associated Milk Producers Inc., the nation's largest dairy cooperative.

Harland and JoAnn Rue

"I guess you could say I was a dairy farmer who worked with cooperatives and that has become my life." he said.

"What a contrast there was in cooperatives. When our 25-member cheese factory held its annual meeting, it was at a school during vacation on a cold day and no one had heated the building. We passed out cigars, hired the cheesemaker and then played cards. But the AMPI meeting in Chicago in the early 1970s filled McCormick Place at a sitdown dinner for thousands and brought President Nixon to the meeting," he said.

Rue's introduction to larger cooperatives that affected public policy started with the old Madison Milk Producers Cooperative, which later merged with AMPI.

But the thing he liked best about the Madison Milk Producers was the association's retirement plan through which farmers could make a contribution with a deduction from their milk checks. The plan grew to an $11 million fund and the Rues will use their portion for retirement travel and special activities.

Rue also was a charter area director of the Wisconsin Milk Marketing Board and its treasurer.

In farming, Rue's corn pacemaker plan made him one of five farms in Iowa County to produce 100 bushels of corn per acre in the early 1950s.

He keeps a close association with farm conditions through his responsibilities as a director of the Barneveld State Bank.

Farming has been rewarding and enjoyable for the Rues and their retirement has been their most difficult decision, Rue said.

Family Builds Two-Farm Operation

June 30, 1991

JUDA — Over the years there have been scores of Illinois farmers who headed north to find a new life in agriculture on Wisconsin farms.

Most of them chose Wisconsin after their Illinois farms were sold either because of urban expansion or investor speculation.

Dave and Vera Kamholz headed north in 1967 with five young children.

But they had no farm to sell because they were farming on shares in dairy, beef, hog and grain operations in the Belvidere area.

Dave said they "had been farming on shares too long and were giving someone else too much. We decided we would be further ahead on our own."

They found their 120-acre farm, packed up their children, a herd of Holstein cattle and moved to Norton Road, off Highway S, in Green County.

They fit right into the life of area.

The Kamholz participation in county affairs is best expressed through Dave's election as president of the Green County Agricultural Chest, a unique organization of the county's agricultural interests.

It promotes Green County's farm products and tourist attractions.

"Not all of our farming life has been a snap," Vera said. "It has taken a lot of sacrifices. We could not have done it without the help of our eight wonderful children."

Three years ago their closest neighbor decided to leave farming and the Kamholz family bought his 156-acre farm. It fits perfectly along the borders of their farm.

Their son Chris and his wife, Darlene, are living on the new farm now that they have joined Dave and Vera in the family farming venture. Another son, Craig, also works with them.

Their Holsteins were replaced with Dave's favorites,

Dave and Vera Kamholz

Aryshires, "the breed I always admired."

He has developed a registered Ayrshire herd with more than 100 cows on the two farms.

The farm is "God's world," Vera said, pointing to the abundant birds, fresh air, and the opportunity to see things grow.

A partially forested bottomland parcel with a flowing stream has been dedicated to wildlife. It is protected from grazing cattle and attracts deer, fox and other creatures.

To take advantage of this peaceful setting, Vera takes a mile and a half walk early each morning.

"As on any farm there is a lot of work to do," Vera said, "but I always tell Dave, if we didn't have all this work to do, we'd be bored."

Dave appreciates farm life, too, because "it has everything, a place for the kids and a place for people to come and be friendly and sociable."

Dave and Vera have remodeled their farm home, landscaped the area and the family collaborated to build a new home for Craig, who was married last fall.

They are members of the Juda United Methodist Church and Dave is chairman of the Juda Future Farmers of America Advisory Council.

Dave also is chairman of an Ayrshire group planning a 1995 National Ayrshire Convention and sale. He also is chairman for the upcoming Ayrshire state sale.

But he really beams when he tells of his job as president of the Ag Chest, which has become a model for other counties whose farmers want to bring farm groups and promotional efforts together.

Dave said Green County created an organization, which is a united way to finance agriculturally related activities to make the best use of human and financial resources.

The nonprofit Ag Chest, Dave explained, was organized on May 1, 1958, when farm associations wanted to organize at the county level.

Each Green County community area, farm and non-farm has a member on the Chest's board of directors, Dave said.

Volunteers are the core of Ag Chest success at such annual events as the Dairy Breakfast on the farm, where 4,836 persons were served this spring.

The Ag Chest will haul its fleet of trailers, Dave said, to all Green County events and sport shows and other activities in cities as far north as Green Bay and south to Rockford and Davenport, Iowa, and the Illinois part of the Quad Cities.

Chinese Officials Enjoy Tour of Top Wisconsin Dairy Farm

January 14, 1996

BELLEVILLE — On one of those wind-whipped, frigid mornings when things don't always go right on a dairy farm, a party of seven Chinese legislators came for a visit to Fischerdale Holstein Farm of Joe and Esther Sarbacker.

On this morning, however, the visitors really liked their introduction to farming and they all personally thanked Joe for the experience and discussions.

It was a place they wanted to visit to complete the Madison part of a six-state U.S. tour, so they could see what it is like to be on a dairy farm, one segment in U.S. agriculture, which ranks as the world's greatest food production system.

During their U.S. visit, which was financed by a grant from the U.S. Information Agency, they concentrated on federal, state and county legislative organizations.

The Sarbackers have lived here since 1961, where they

Joe and Esther Sarbacker

not only raised great dairy cattle, but also 11 children, after starting in dairying on other nearby farms. Joe is grateful that all seven sons are still in dairy agriculture and four daughters are still associated with agriculture.

The Chinese visitors were brought to Dane County by the National Committee on U.S.-China Relations. They are just the most recent of many international visitors the Sarbackers have welcomed. Last summer, they were joined by a young man from the Netherlands.

The seven Chinese visitors started their Sarbacker tour in the barn, which was welcome relief from the cutting winds outside, since none of them wore a cap or hat and all were in business suits and light coats.

The reward for touring the barn came 45 minutes later when they were invited to Sarbackers' home to sit around the dining room table, filled with cheese plates and delights prepared in Esther's kitchen.

They wanted to hear Joe's ideas on farm economics and the cooperative organizations which make him a part owner of the 10 co-op businesses he depends on to successfully operate the farm of Fischerdale Holsteins.

"They seemed reluctant to leave and I thought they wanted to talk all day," Joe said. But the time flew by and they had to be reminded of their 1 p.m. flight from Madison.

After Joe's introduction to the farm was translated, the Chinese asked questions and a skilled American interpreter passed on Joe's remarks.

When Joe listed the number of cooperatives he relied on in his farm business, it had special meaning since the cooperatives are business units in which he is a part owner. Questions came back to Joe on his part ownership as a patron in each cooperative business. The guests wanted to know the number of farmers in a cooperative and asked how Joe was represented in the organization.

The member-owners elect their board of director members in regional meetings. Joe knows the business because he has served as a director in a few of these units through his election by fellow members of each cooperative.

In the farm office, the Chinese admired the full shelves of gold and silver trophies, plaques and other prominent awards which Joe and his family had won with their great line of cattle.

Joe handed several guests a trophy to hold so they would know what it felt like to be a winner. The Chinese were all smiles.

Mr. Knigge Goes to Washington

April 8, 1990

Probably no one in Wisconsin agriculture has drawn more attention in the past two years than Pete Knigge. David Ebert, vice president of Valley Bank of Oregon, referred to Knigge "as the man from the Omro and Washington, D.C., area" while introducing him at Valley Bank's recent session on agriculture.

Knigge quickly responded: "I did the milking at home this morning and if I get out of Madison soon, I'll also do tonight's milking."

Knigge, whose real first name is Charles, does his milking — when time allows — on a 240-acre farm near Omro in Winnebago County.

When he's away, as he's frequently been for the past two years, he's likely to be in Washington, where he chairs the National Commission on Farm and Rural Development. It was created by Congress in the 1985 Farm Bill.

When Pete is on a trip, his wife, Theo, and his children handle the dairy operation. The Knigge family — Pete included — is doing an excellent job on the farm.

Their Guernsey herd ranks fifth in the state in milk production at 15,000 pounds of milk and 745 pounds of butterfat per cow per year.

Knigge is especially proud of the herd's milk quality as measured by a low somatic cell count of 75,000. Knigge said he wishes he could be as effective in getting members of Congress and consumers throughout the nation to fully appreciate what agriculture and farmers mean to them.

He also worries that farmers have a reduced role in influencing federal farm legislation. Evironmentalists have become stronger, he said, and "they will have more input into the 1990 Farm Bill than some of the major farm organizations in the country."

Knigge said environmentalists are adamant in their demands and back that with political clout. "You have to realize there are 10,000 more people paying dues to environmental groups than there are farmers in the United States. That's the basis of their political clout," he said.

Knigge also has reservations about the nation's cheap food policy and said agriculture is taking a bum rap from consumers over the cost of federal farm programs.

Pete Knigge

"I guarantee you (consumers) are going to continue to complain about cost of the farm bill even though the costs continue to decline," he said. "That's the way it is going to stay when 97 percent of the people are food consumers and 3 percent are food producers."

Knigge's job as chairman is to oversee the commission's hearings, visits and reports. The work is voluntary — he receives only expenses.

Still, it is a position requiring experience, success in farming and the ability to clearly convey farmers' messages.

Knigge is confident in his abilities, having found that it pays to submit a resume when a person is convinced he or she wants to take a position in government.

He did so when he learned of an opening on the State Board of Agriculture. He did not get the state appointment, but in 1988 Gov. Tommy Thompson used the resume to nominate him as a member of the national farm commission. President Reagan appointed Knigge and the commission's members elected him chairman.

Thanks to UW-Madison, Knigge is well trained for his job, having graduated from its dairy science and rural leadership programs.

"The UW's two-year (Wisconsin) Rural Leadership Program is right in my line because it has been a stepping stone for me," Knigge said.

"I personally wanted to learn more about what is happening in agriculture and rural living. The leadership program provided that. I want to be a spokesman for agriculture."

State's Top Farmers Love their Work

March 31, 1991

DODGEVILLE — The brilliant flairs of sunshine breaking through the irregular March clouds flooded the kitchen in the McGraw farm home like an Easter morning spotlight centered on the couple seated at the table.

And it was the way it should be because Dennis, 34, and Sue McGraw are something special in Wisconsin.

Together they share the honor of holding Wisconsin's Outstanding Young Farmers award for 1991. From a selection of the state's top young farmers, the McGraws represent the generation of young people in farming who will lead Wisconsin agriculture well into the next century.

They were sent to the state finals by the Dodgeville Jaycees and will compete in the national OYF contest 11 months from now in Moline, Ill.

Their prizes include use of an IH Magnum tractor for 100 hours, and a utility tractor and Gehl manure spreader, and a $250 gift of semen.

They are dairy farmers who, like their neighbors in the past five years, have been through one of the most demanding tests any generation of farmers has faced.

Dennis noted the drought years of 1988 and '89.

But they emerged from the seige of dry weather, had good crops in 1990, gained record milk prices for 1989, only to face a sharp collapse of milk prices in the last five months of 1990 and at a lesser rate in 1991.

But here they are, young, resourceful and confident their two sons, Matthew, 9, and Peter, 5, and their daughter, Katie, 2 1/2, will be farming one day if that's their decision.

"Matt and Peter are talking farming," Dennis said, "and Katie, is just enjoying life behind that big smile and apparently hasn't made up her mind."

But to the outstanding McGraws farming is still "fun." They are playing it tight right now because of the adverse dairy economy, but still place their hopes in their 68 milk cows and a small feedlot for dairy steers.

They are doing an exceptional job with the herd since the milk production average per cow is 22,000 pounds — 8,000 pounds over the state average.

The herd is led by a cow their son, Matt, won in a Tri-State Breeders drawing when it was a heifer at World Dairy Expo. The cow's production record for 305 days is 31,000 pounds of milk.

While Dennis grew up on the farm of his parents, Paul and Lois McGraw, four miles east of Dodgeville on highway 191, Sue came to farming after living as a city girl in St. Paul.

David and Sue met and fell in love while both were students at UW-River Falls, where Dennis was majoring in agriculture education, and Sue was majoring in elementary education.

But when Dennis got a job as vocational agriculture instructor for the 1979-80 school year in Muscoda, there was only one unanimous decision — they got married the same week Dennis graduated.

Together they'd face agriculture, first from the classroom (only a year), and then from the McGraw farm, where they began farming in 1980, and farmed at another McGraw farm until they purchased their present 370-acre farm from his parents in 1985. They rent an additional 120 acres.

His father retired in 1985, but he still comes out to lend a hand almost daily

Dennis said he offered to come back to farm when his father purchased another farm, one he heard about in church one Sunday.

"I liked teaching agriculture, but I guess farming was my first love," he said, until Sue raised her eyebrows. "Second love," he said, correcting himself.

And guess what? Last week they purchased a computer.

How fortunate they are. Sue's father, who lives in St. Paul, is a retired state computer expert and ready to give them a course in computer operation.

Who is going to enter all the data on farm operations, milk production per cow, crop yields and financial records and costs of operations?

"I guess I'm going to do it," Sue said — slowly. One of the farming advantages the couple has, is that two other McGraw farms are nearby.

"With the three McGraw farms we can share farm equipment. It is so nice to work together, sharing help, too," Dennis said.

The McGraws both are active in community and church affairs. Sue is on the education commission of St. Joseph's church, is chairwoman of the monthly fish fry, and does volunteer work in the school office.

Dennis is a member of Farm Bureau, a Tri-State Breeders Cooperative delegate, member of Associated Milk Producers Inc., the Dodgeville Jaycees and the Dodgeville High School agriculure department.

He said the county Extension office, feed mill and Dairy Herd Improvement Association have the resources available for his operation when he needs help.

"We love farming and it is a career choice and a business, too — big business and has to be run that way," Dennis said.

He Stuck with Farming Through Thick and Thin

March 3, 1991

COOKSVILLE — In the little red brick house on Webster Road that easily could have been the House of the Week back in 1850, Joseph Franklin Bradley sits in a small office decked out with pictures and documents of his 77 years of family, farming and community.

If you wonder who Joseph Franklin Bradley is, you should understand it was the state of Wisconsin that altered his name. This happened when one of the state's first electronic registration systems wouldn't accept his preferred name of J. Franklin Bradley.

"The person at the drivers licensing counter said the machine couldn't accept "J" as a first name so I no longer was Joseph," he chuckled.

Now Franklin Bradley spends hours in his well-organized office — warmed by a blue electric heater with a steady grumble, which often drowns out Bradley's subdued voice.

Bradley was born on a farm next to Wind Point in Racine. After graduation from the Racine County School of Agriculture he went to Madison and the University of Wisconsin College of Agriculture, and joined the Class of '37.

Bradley, who lived at Babcock House, 432 N. Lake St., once the home of the famed UW scientist, still gets a laugh when he recalls how Madison's newspapers reported a complaint that "the boys at Babcock House were seen playing baseball in Babcock's beautiful hollyhocks." Babcock, who developed the test for butterfat in milk, was a hollyhock fancier.

Bradley majored in animal husbandry and agricultural engineering and graduated in 1937. The "W" he earned was for his service as manager of the UW basketball teams under Harold E. (Bud) Foster as the coach built up the team for the 1941 run for the NCAA championship.

After his marriage in 1938 to Ardith Hurley, daughter of R.V. Hurley, one of the first Dane County Extension agents, he put both his agricultural interests to work on the farms he either worked on or owned in Racine, Kenosha and Rock counties for the next 45 years.

He retained his interest in farm machinery and equipment and still is a serious student of the technology of field equipment. Today he represents Badger and Berg equipment companies.

Rock County has been his base since 1946 when he moved to the area east of Janesville where he operated the 750-acre Stone Farm until 1950.

That's when he and Ardith became owners of Ardlin Farm, a dairy operation (Guernseys). He and his family of three children operated the farm until 1983.

One of the gems in his home today is a quilt with squares that portray the history of the family of Ardith and Franklin. For those who were too young to prepare their segments, the squares were their drawings on the fabric with crayons.

Having his own, adequately sized farm was a goal Bradley always had after a trusted, experienced friend advised him, "it's foolish to work for anyone else. Be independent and just stick with farming."

And Bradley had the independent spirit needed in agriculture, even when he took nearly a $100,000 loss after the Ben Green's grain service went under. He recovered about a dollar for each $1,000 he lost.

In every niche of his office and around his fine home (even with the leaky flat roof), Bradley has some little catch phrase framed or pinned to the wall.

For instance, he has one brief message reminding visitors that "the taxpayer is a person who doesn't have to pass a Civil Service examination to work for the government."

Or, "patience is idling your motor when you feel like stripping your gears."

And who can disagree about the finding that "an optimist is wrong as often as a pessimist, only he is happier." What must have been the first anti-litter message, was one Bradley found at Camp Upham Woods, the 4-H camp at Wisconsin Dells: "If you with litter would disgrace and sad this place, may indigestion wrack your chest and ants invade your pants and vest."

Bradley found single life difficult after the death of Ardith, so he is planning a spring wedding, but not revealing the name of the bride so he can surprise his friends.

After the wedding, they plan lots of travel in the newly acquired travel home.

"Life has been good to me," Bradley said.

Arnesons Take Farming Expertise to Washington, D.C.

May 7, 1995

STOUGHTON — Darlene Arneson has quite a job to do as president of the Dane County Farm Bureau, since there are times like a few weeks ago, when she represented the organization's 635 farmer voting members and took their 1995 Farm Bill message to Washington.

Her trip, along with her husband, John, included visits with most of the Wisconsin congressional delegation. She also met with U.S. Sen. Richard Lugar, R-Ind., a presidential candidate and chairman of the Senate Agriculture

Committee.

Darlene's big surprise, however, came when she was invited to the Agriculture Department, where hundreds of department employees were meeting Dan Glickman, the new U.S. Secretary of Agriculture, less than a day after he took his oath of office.

There were many couples from Wisconsin on the Washington trip — all of whom paid their own expenses — and Darlene said the trip was successful even if the 1995 Farm Bill will not be written until later this spring and summer.

It is possible that the Wisconsin Farm Bureau will organize a second trip this fall when the farm bill is nearing completion or even ready for a vote.

As Dane County president, Darlene takes her responsibilities seriously and knows the detail, background and reason for basic dairy and specific farm policies.

Darlene and John are dairy farmers and Darlene can discuss dairy policy at length.

She was trained for her county leadership post by serving three terms on the county board of directors before being elected president.

Darlene also gained leadership ability from her FFA experiences at Verona High School, her farm experiences and her UW-Madison education and election to a state FFA office.

She gained additional experience through FFA by spending time on farms in Norway, Ireland and Australia. This preparation made her ready for a vocational agriculture teaching career, including a total of four years at schools in Cashton and Cambridge, plus two years of preparatory work at Rio.

John adds to her experience with his state FFA service as president and vice president.

Today Darlene and John have a unique farming partnership with Eric and Marge Stokstad. At the farm on County

Highway B, north and east of Stoughton, the Arnesons have the dairy herd and manage the dairy, while the milk goes to the new cooperative enterprise, Foremost Farms.

The Stokstads own the farm and the farm equipment. The Stokstads have made regular barn and feeding improvements. The barn now has 51 stanchions and John says, "We keep them full all the time."

Young Families Form State's Farm Future

November 20, 1994

BELLEVILLE — Lori and Doug Brown — and their two daughters, Katie, 9, and Carlie, 4, — are part of the new generation of Wisconsin dairy families — a generation that state and local leaders want to see prosper to maintain the state's agricultural base.

They are starting in a dairy farm partnership with Doug's father and mother, Jerry and Jeanette, on the family farm on Highway A, three miles north of Belleville.

The important part for Lori and Doug is that with their skills as young farmers, they are being recognized, encouraged and honored by organizations such as their dairy cooperative, Wisconsin's Morning Glory Region of Associated Milk Producers. They also were cited as Outstanding Agriculturists by the Downtown Kiwanis Club of Madison.

Soon, they'll be in Dallas to join other winners of the Outstanding Young Dairy Farmer Award from around the nation at the National Milk Producers Federation meeting.

Fortunately for Wisconsin, which is on a steady milk production rebound after two difficult years of poor weather and poor crops, young farmers have new reasons to feel confident.

It is because of things like the 176-bushel-per-acre yield of corn Brown harvested and the high tonnage of alfalfa hay — from four crops this year — which to Doug mean a tremendous saving in the cost of purchased feeds for the farm's 116-cow milking herd.

In addition, the recent four consecutive months of milk price gains were especially encouraging given the declines in livestock and grain prices.

The Browns say if they weren't optimistic about dairying, "we wouldn't be here." In addition, Lori and Doug credit their family for giving them the spirit and dedication to succeed. An example is their daughter Katie, a fourth grader at Belleville elementary school. She comes home by bus, sits down in the house and does her homework and then, tired as she may be, runs to the barn to join her parents as they finish the milking.

"She's super," Lori beamed. And she's hoping for another daughter when she adds a new member to the family early next year.

Each Wisconsin county is dotted with young farm families filled with hope, just like the Browns. Those family farmers make up the potential of Wisconsin's dairy industry.

Lori and Doug met at the Dane County Fair when showing their Holstein cattle year after year. Lorie was a member of the Verona Go-Getters 4-H club, while Doug was with the Belleville Busy Beavers club.

"We were competing then. When the cattle judge finished the class he would have placings that put us side by side, like first and second," Lori said with smile, "And we are still competing."

The Browns, in their early 30s, were married 10 years ago. After a short trip to Niagara Falls, N.Y., they quickly returned home because it was corn planting time.

Dairy Family Struggles with Chores

March 27, 1994

On Route 2, Soldiers Grove, Bill and Lynn Kloehn operate a farm with an 80-cow herd of Holsteins, 80 head of young stock and a dairy-beef feeding operation.

Things are going well, particularly with the 80 cows having Dairy Herd Improvement milk production tests showing a rolling herd average of 20,000 pounds of milk per cow annually. (That's about 6,000 pounds over the state average.)

In addition, they regularly market 1,400-pound Holstein (beef) steers raised and fed from bull calves in the home herd.

But there are times, Lynn quickly adds, that it takes three generations to keep the farm going: her father, Gilford Dodds, the farm owner and senior generation; Lynn and Bill Kloehn, young farm couple carry out day-to-day operations and working on a share arrangement so one day the family farm will be theirs; and — the third generation — the Kloehns' growing family, Michael, 10; Joe, 8; and Becky, 6.

Lynn, a part-time nurses aide, does double duty, mostly working night shifts at the Sannes Skoldalen Nursing Home so she can be home at those critical moments making hay or silage in summer and fall. She also is there many times to help with the milking.

Bill says with everyone working together they are making progress, "but we recognize that a successful operation can make only minimal mistakes or you are out the door."

Bill grew up in the Hub City-Yuba area of Richland County, where his dad had a dairy and beef feeding operation.

With his father as a teacher, Bill learned enough about a dairy and dairy-beef operation to be successful today. He is a graduate of Richland Center High School and had some well-rounded agricultural programs as a member of the school's Future Famers of America chapter.

But there still are those days of stress that come to many farm operators. When that happens on their farm, Lynn said, "we just get away a day or two and get recharged."

They were in Stevens Point last weekend as Richland Center Equity Market representatives to the annual Young Farmers Seminar conducted by Equity Cooperative Livestock Sales Association of Baraboo.

Bill said it gave the 46 young couples a chance to know just what a variety of programs Equity has, especially in marketing and feeding, including byproduct feeds.

Lynn said it was worth the time investment and she hopes other young farmers using the Equity systems apply for the seminar program. She was impressed with the ideas offered by the seminar students and suggests other young farmers offer their ideas to the 46 couples attending the sessions.

The current issue among dairy farmers — whether recombinant BGH should be used in their herds — came up for discussion.

Young farmers were asked to indicate how they felt about recombinant BGH and about half of them indicated they did not know whether it was good or bad.

"I haven't explored it fully yet," Bill said. "Out of my herd of 80 cows, I think I have only four or five cows on which I would use it," he said.

"These are cows that can work (in producing more milk) by getting more fat off their bodies. Otherwise they just stay fat and lazy," he explained.

"If I can use rBGH and increase milk production to make more money, that's the route we should go. That's why we are in business," he added.

Bill said he recognizes the political moves being made to discourage dairy farmers from using rBGH. His milk is shipped to Springdale Creamery, Route 1, Richland Center, a Muenster cheese operation, which has 165 patrons and is seeking more quality milk.

He said farms in his area with larger herds have been selling off cows and now these same farms are buying cows to get back into full production.

To bring milk prices in line with city wages, Bill believes the average all-milk price to dairy farmers should be about $17, compared to $13 now, for the combined use of manufacturing and fluid milk.

Holiday Spirit Survives Hard Times

December 24, 1989

BOSCOBEL — With the arrival of Christmas, you may think back through the past year and find things that did not go the way you and your family hoped.

Everyone has troubles of various degrees, but if you should ever sit at the kitchen table in the farm home of Gloria and Dennis Brandes and their four children to talk about their year as a family, chances are you would head home believing you and your family are fortunate.

The Brandes family didn't have much luck over the past seven months, but they did have determination, family love, and kind and supportive neighbors and friends.

Ultimately the six Brandes showed their willingness to prevail despite unbelievable adversity.

They are going to stay in farming despite the worst

combination of circumstances from family health to a setback of a ruinous barn fire.

The Brandes have a 198-acre family dairy farm that currently falls short of that description. The 45 dairy cows are 10 miles away in another barn while still another barn is being fixed to handle the herd. Gloria sat at the head of the table while a quiet Dennis sat for a time, listened, then quietly disappeared into the ridge-top cold to keep Sun Rise Farm going. The Crawford County farm is on Plainview Road off Highway E, seven miles northwest of Boscobel.

It's been a hard year for the family — Dennis, Gloria and their children, Lori, 18, Lisa, 17, Jonathon, 14, and Aimee, 11.

Back in May Dennis hurt his back and was laid up a week. About the same time, Gloria, ruptured a disk in her back but unlike her husband, this injury was not resolved in a week's time.

Those conditions were bad enough, but things started to compound when Lisa, helping out with the barn chores was kicked by a cow and injured.

Jonathon was also injured in May playing tether ball when he jumped and landed in a hole, breaking both bones in his left leg.

In June, Lori went to the hosptial to have her appendix removed.

Ten days later, her sister, Aimee, was in the hospital for the same reason.

What about July?

Gloria was in the Boscobel hospital for 10 days before being transferred to Madison for kidney stones. In August, Lori had two surgeries to correct an abdominal infection.

Aimee remembers an August bicycle ride when she fell and broke her arm.

But the worst was to come — on Dec. 1 the 32 by 114-foot barn burned with the loss estimated between $80,000

Gloria and Dennis Brandes

and $100,000.

What's worse, a month earlier their insurance company cancelled all insurance including fire coverage when a $9 added premium was not paid on a piece of haying equipment.

The main insurance premium had been paid, but when the insurer called to say the add-on equipment premium had to be paid or "it" would be cancelled, Gloria thought the cancellation applied only to the added haying equipment coverage, not the entire policy.

The matter is being investigated by the State Insurance Commissioner, with the help of State Rep. Dwayne Johnsrud, R-Eastman. The commissioner's report will be completed in January, Gloria said. They have not hired a lawyer, she said, because they do not have the money.

The Brandes did, however, have health insurance through Associated Milk Producers Inc., their milk marketing cooperative.

There is still the spirit of Christmas at the Brandes home, where the Christmas tree is up and the house decorated. The spirit was helped by the many people who contributed to a benefit auction for the family.

'Its the Greatest Place in the World'

December 1, 1991

The Voegeli Farm on Highway 69, just west of Monticello, has a solid future, even during these difficult times on a dairy farm.

Howard and Alice Voegeli are the family's fourth generation and their two sons, Bryan and Jimmy, make up the fifth generation of the distinguished pioneer family of Brown Swiss cattle breeders. The Voegeli's have two daughters, Jacqueline and Rhonda.

The Madison Downtown Kiwanis Club used this year's Farm-City Week as the occasion to present its annual Agricultural Service Award to Howard and Alice for their farming and Brown Swiss cattle breeding ability as well as for their involvement with rural youth and their volunteer service.

Their picturesque farm in the Green County hills with its fine farm home, massive white barn and a heifer shed made of oak cut from the 300-acre farm has changed drastically since it was settled in 1854. The Voegelis rent an additional 450 acres for their crop and livestock needs.

Voegeli recalls how he came back to the farm after finishing his UW-Madison studies in dairy science as his classmates looked forward to their new careers and new jobs.

"But I had to go home because my dad, (Jake Voegeli) was leaning on me to help on the farm. I was not excited about doing this," Voegeli said.

Voegeli started with all the vigor of youth, milking his big Brown Swiss herd three times a day to get more and better production records.

"I used to get up at 2 a.m. and work to 7:30 p.m. I didn't know what day of the week it was, but that didn't matter — they all were the same, anyway," he recalled.

And then Alice Lien, Blanchardville, soon to be Alice Voegeli, came into Voegeli's life and "she changed all my views," he said.

"We struggled to buy into the farm and many times we sat at the breakfast table wondering how we were going to make it," he said.

But almost on cue some foreign buyer would come in and buy a lot of cattle because he was looking for milk production records. Voegeli found that milking three times a day paid off.

"We were excited as we watched the truck pull away with a load for Cuba or Venezuela, our markets in those days. As we watched the truck disappear we knew we could survive for another year. Thank God for Brown Swiss cattle," Voegeli said.

Now the farm has organized Voegeli International and has sold Brown Swiss cattle to dairy interests on four continents.

Together they were able to improve the farm operation. And, Voegeli paid tribute to his wife, Alice: "Without Alice, Voegeli farm would have been closed a long time ago," he said.

"Every time I needed someone to help milk the cows, she was there, even to milk by hand. She's a tractor driver, calf feeder, reminder of things we have to do, goes for things we suddenly need. And most of all, she's our accountant, the payer of bills," he said.

"After working in the field with us she would run into the house, he continued, and make a meal fit for a king. Look at me now, you know what I mean," he laughed.

Alice Voegeli also sings with a trio and entertains at nursing homes, meetings and other groups. On Sunday, Voegeli said, he has no hired labor, so the family milks

Howard and Alice Voegeli

the 100 cows.

Voegeli then tells what farm life is all about: "I take a little brandy and go out and sit at the picnic table thinking of things that happened last week and what's going to happen on these green hills with cattle on the grass.

"I'm going to stay here. It's the greatest place in the world. Come and visit us sometime. I'll show you why."

They Stand in a Long Line of Farmers

August 11, 1991

WEST ALLIS — The Lehmans — Connie and Gerald — patiently stood at the end of a long breakfast serving line and even though the couple doesn't like standing in lines, this was one they wanted to be in.

The line at the Wisconsin State Fair included individuals and family members who were going to be honorees as owners of the new class of Wisconsin "century" farms and homes.

Connie and Gerald Lehman

The Lehmans, who farm enthusiastically and successfully in the Town of Ironton, near Reedsburg, were saluted by the state for having their farm continue with the Lehman name since it was acquired by Jerry's great-grandfather in 1881.

It also was the oldest property in this year's Sauk County list of property owners. Other properties included Melvin and Loretta Bickford, Prairie du Sac, with a farm going back to 1887, Willard Brechtl, Plain, back to 1891, and Marvin and LaVerne Giebel, Reedsburg, back to 1890.

The 176-acre Lehman home farm is about double the

size it was when the Lehman name was first put on the land. Naturally, the farm carries a lot of family history — Jerry and his father were both born there.

Jerry's research showed that when the farm was purchased 110 years ago, it was another tough time in agriculture. Because of the economy then, Jerry said, his grandfather got more than just land when he bought the property.

The bill of sale, he said, showed that along with the farm's original 80 acres, his grandfather received four horses, three yearling colts, two foals, six cows, four heifers, six sheep ("the only time sheep were kept on a Lehman farm") a few chickens and a plow — all for $2,500.

"So you can see how the depression at that time deflated land values, just as it did in 1900 and in the 1930s and the recession of the 1980s," he continued.

Today's $2,500 value could be reflected in the sale value of one of Jerry's top Holsteins brood cows in the dairy herd.

Connie became part of Lehman farm history when she married Jerry 23 years ago. She grew up on a dairy farm in the Sauk-Prairie area and met Jerry in 4-H clubwork. They married after her graduation from UW-Madison.

Today Connie is a farming partner with Jerry and keeps the books and does her share of chores in addition to teaching mathematics at Reedsburg's Webb High School.

And chances are good that the tradition they've built upon will continue. Jerry and Connie have two sons, Mark, 18, and Kurt, 13.

The Lehmans have a 65-cow dairy herd on the home farm, which accounts for a million pounds of Grade A milk being shipped out annually to the AMPI Hillsboro muenster cheese plant.

They also operate a dairy beef feedlot on a rented farm to feed 35 Holstein steer calves annually. Jerry said that although the Holsteins produce milk, "We also operate the herd as kind of a brood cow herd because every calf is going to make some meat some day, either as cull cow or a Holstein steer.

"We realize that as dairy farmers we also are in the meat business, which is something many dairy farmers don't realize," Jerry said.

"We have 290 head of cattle, including the dairy herd and a registered Shorthorn herd and a crossbred beef herd on six different farms," he continued.

Jerry is a member of the National Cattlemen's Association and when the beef cattle leaders get crossways with dairy policy and attempt to block and or defeat dairy policy, he is one of the first to help beef leaders know the facts of dairy policy.

Jerry believes dairy farmers as beef producers could have the numbers to help beef people adopt a good, fair dairy policy.

"The problem is that dairy farmers with NCA memberships don't attend the meetings where they could influence policies of national cattlemen," he said.

"Remember, you have to join them to argue with them. You can't just sit on the sidelines and complain as the beef contingent knocks down dairy policy," he said.

He also is a forceful, logical person when appearing at hearings of the Wisconsin Legislature and often has fought some tough fights.

"There is nothing I like better than going to a hearing. But the committee members always want to know who I am with and who I represent," he said.

"I represent myself and that's something they don't understand," he added.

Jerry stays extremely well read on dairy policy issues, which are currently being debated in Washington, as well as beef and environmental issues affecting agriculture. He now serves as chairman of the Town of Ironton.

Farming Goes Back 100 Years for Gants

March 7, 1993

BLANCHARDVILLE — It's spring, really spring out here on the Gant Farm.

Regardless of what any calendar may say, the season arrives when it's calving time.

Karl Gant

Those good mother cows in the Hereford herd, who spread the natural technology to the first calf heifers, are again bringing new life to the farm and getting everyone like Karl Gant focused on what farm life is all about.

"I have 120 cows to calve and I've had 16 calves so far and one lost. I think she was premature," Karl reported. It has been a good calving season thus far, except for the one heavy snow on the Sunday a few weeks ago as the season was just getting started.

Gant, 48 and unmarried, is a historian, a guardian of nature, a man of God who fills in for the vacationing Methodist pastors and once served three churches, a sales manager for the South West Wisconsin Feeder Cattle Cooperative for 14 years, and a dedicated independent beef farmer.

He is one of those rare persons who, like his father, 91-year-old Bernard, and his brother Jerry, has never used a government farm program.

There is one other focus Karl has and that is taking care of his mother, Lulu, 85, who is in an advanced stage of Alzheimer's disease. She hasn't communicated with anyone for the past two months.

His brother, Jerry, and sister, Barbara, who lives in Stoughton, help a lot. He has two women to stop in and clean.

"But our mother would do the same for us, too, if we had similar problems," Karl explained.

All these things happening on the Gant Farm, a century farm, are part of the scheme of things on the 1,950 acres his great grandfather, Jim Gant, started putting together when he bought the first 40 acres in 1889 to acquire grazing land and raise beef.

The family traces back to Southern England and the entire family of 14 came to America together. Jim worked as a cattle buyer at Jonesdale and shipped market-ready cattle to Chicago by rail.

Karl said when Jim bought land he picked his grazing land carefully to make sure he had one stream and a spring in each 40 acres.

Except for one 33-acre parcel in Iowa County, all of the Gant Farm is in northern Lafayette County. Karl bought land of his own recently. He acquired the acreage as recreational land so he could get away from things to restore his spirits and maintain the land.

He will say he never sees any deer on his property, but his friends know the deer will run and follow him and of course he doesn't see deer because he doesn't turn back to take a look. He communicates, too, and the fawn lying in the grass will simply get up and slowly walk away.

This land for a time was owned by Boise-Cascade, but the development plan never worked and land not purchased by Karl was subsequently bought by the Department of Natural Resources as a wildlife area.

Karl recalls how he was urged to plant 80 acres of his purchase to trees as part of the government's 10-year Conservation Reserve Program, but he steadfastly resisted even though the acres would have earned an annual government payment of $5,600.

That's just an example of the thousands and thousands of dollars in federal farm funds the Gant family never used because they were in farming for the challenge and were willing to accept it.

Three-day Bash for Longtime Dairyman

January 26, 1992

OREGON — When Reginald Maxwell Gefke celebrated his 85th birthday recently, it was a gala event that stretched for three days.

"And all those cakes and cards," he said with a modest smile.

The birthday balloons, floating in his living room, all have appropriate cow portraits and dairy messages, such as: "just be-cows you're special."

He lives in retirement after more than 60 years in dairying, but he can be seen at every dairy gathering because he loves the association with the new generations of dairy people.

A table in his neat and tidy home is covered with colorful cards from friends.

But what is this Reginald Maxwell Gefke bit? After all, he is a man who has been known as Max all his life.

R. Maxwell has been his official name since he registered for Social Security 20 years ago and officials said they would only accept his full name. His first name, Reginald, he explained, came when he was named after an early Madison physician, Dr. Reginald Jackson.

Gefke has been living on his farm on the east edge of Oregon for 84 of his 85 years. He was born in 1906 on a dairy farm on Highway 21 between Tomah and Necedah

R. Maxwell Gefke

and a short time later his family moved to Oregon.

In recent years, the original 220 acre farm, which had been operated by Max and his brother, Bob, was greatly reduced in size as part of it became an Oregon golf course and 70 acres was needed for construction of new Highway 14.

Max tells of a big milk route the farm had from 1929 to 1951 to serve up to 200 Oregon consumers with milk bottled daily on the farm. When state pasteurization rules were adopted in 1951, the route was sold to Bancroft Dairy of Madison.

Earlier his grandfather, Joseph Gefke, took the milk to Oregon in a large tank. He was something like today's good humor man and he would ring a bell as he went down each street. The customers would come out with pails in which to carry the milk dipped from the tank.

Gefke vividly remembers the time the farm acquired its first registered Holstein cow in 1914 and how the herd was built to a 60-cow registered herd that was regarded as "big" in those early days.

Events on the Gefke farm were troubled in 1914 when a neighboring farm had an outbreak of hoof and mouth disease traced to the importation of feeder lambs. The neighbor lost his entire herd.

There was a state inspector on the Gefke farm every day, but fortunately the herd never was infected with the dreaded disease. On farms where two herds of cattle were infected, deep trenches were dug in the fields, the cattle were herded into a trenches, shot and buried.

You still can see evidence of those trenches on lands west of Oregon, Gefke said.

The Gefke dairy herd was milked by hand until the 1930s, when milking machines were first developed. Gefke didn't like the first model and went back to hand milking until a better machine was introduced.

He graduated from Oregon High School, where sports were a major interest since he was on the first Oregon basketball team to get to the state tournament.

Gefke then enrolled in agriculture at UW-Madison, where he was graduated in 1928.

He will never forget that graduation ceremony. It was to be held in Camp Randall stadium, but when it appeared that rain would dampen the ceremony, it was shifted to the stock pavilion on the agriculture campus.

By the time the graduates got there, all the seats were filled and the graduates had to wait outside and entered only to walk across the stage and receive their diplomas.

He was offered jobs in dairying, but he decided to stay in Oregon — "a good community," where he became active in various farm and community organizations. Today, he is treasurer of the Oregon Historical Society. His contributions of old equipment to the museum include a John Deere wood beam plow, one of only 12 such models made.

Gefke was married to the former Thelma Nelson, a teacher and champion pie-maker.

Gefke's active days on the farm were long — from 5 a.m. to 8:30 p.m.- but he never regretted his decision to stay on the farm. Before there were cattle trucks, Gefke rode the train to take his show cattle to meet the competition as far away as St. Louis.

In 1985, when Gefke was 77 years old, he quit dairying. "I always liked dairying and I still do. I like cows, farming and everything that went with it," he said.

Chapter 2

Dairy Cooperatives

Torgerson's Book Relates History of Cooperative

May 6, 1990

The UW-Madison hockey team celebrated its championship this spring at the same time an earlier UW NCAA champion celebrated his special achievement by writing a book.

Truman Torgerson, Manitowoc, won the light heavyweight NCAA boxing crown for the UW in 1939 when he defeated a Syracuse University knockout king.

He has written a superb people story of the beginnings and life of a dairy cooperative and how farmers, with great leadership and motivation, were able to work together and succeed in the business world.

His 352-page book, "Building Markets and People Cooperatively — The Lake to Lake Story" reflects Torgerson's 35 years as president and general manager of dairy cooperative Lake to Lake.

In 1980, the 35 years of Lake to Lake as a separate dairy cooperative ended as it was merged with Land O'Lakes Inc. and became a division of the large Minneapolis-based regional food cooperative.

The book is especially significant since Torgerson wrote as the only general manager the cooperative ever had. The story also is about how the University of Wisconsin can help people who ask for its expertise to improve their lives and fortunes by forming a cooperative and taking it to the top as a national dairy production and marketing association.

The book is a tribute to the late Rudolph Froker, an

agricultural economist and later UW dean of agriculture, who developed the concept of a large dairy cooperative in the area from Lake Winnebago to Lake Michigan.

Also, Torgerson credited the contribution of Professor Walter Price, the university's cheese expert, who helped make Lake to Lake's cheese a national product when it became the first U.S. cheese manufacturer to qualify for the U.S. Department of Agriculture's Grade AA Shield.

The book also is a lesson in business management that can apply to any business, proprietary or cooperative, that is working for the good of people and the community.

But most of all, Torgerson's historical effort is a book for the many people of Wisconsin who never have had the chance to know and appreciate what a cooperative is in the business world.

Truman Torgerson

Torgerson's writing style is like the champion boxer he was for Wisconsin — he had the punch and he used it. "We tried to be honest all the way through," he said. "We can't do enough cooperative education. I'll tell you that."

Torgerson had a way of working with farmers and their wives, the kind of home farm team needed to move the assocation effectively and efficiently.

He gives great credit to farm wives as partners in the farm operation and what they meant to the cooperative. "Once the farm wife was sold on the cooperative concept you could more easily sell the idea to the husband," he said. Much of the book tells how to motivate people

— Torgerson style — with education, goals and pride of being part of an organization that is going somewhere for and with people.

"When members take hold of their own affairs and when you get them in the right frame of mind you've got to educate them, guide them a little bit ahead of time," he continued.

That's because this Norwegian farm boy from Dunn County early in life found the cooperative organization as something special for people.

Torgerson was a Depression person who grew up on the farm having nothing. At the same time livestock shipping associations were being organized by farmers. He heard speakers urging farmers to help themselves through their own groups.

"To me cooperatives were a way for people to get ahead. I always felt I should try to help the farmer because his income was so bad all those (Depression) years. I never forgot it," he said.

He had good teachers, first at the Dunn County School of Agriculture where Randall Swanson, "my ag teacher" and now a UW emeritus professor, urged him to go into agricultural education and county agent work at the UW because "that will prepare you for anything."

Torgerson described Professors Asher Hobson and Marvin Schaars of agricultural economics as "great teachers."

After graduation he was an agriculture teacher at Tony high school and served as county agent in Rusk County and then in Manitowoc County. While at Tony he coached the boxing team and the boxing matches were so popular they drew more fans than the basketball team.

Don Storhoff Helped Lay Foundation for Strong Dairy Industry

October 15, 1995

Don Storhoff, president of Foremost Farms USA, is a true dairy business executive, leading a movc to improve the dairy economy and business climate for the dairy farmers and their industry in the Upper Midwest.

From Foremost Farms' headquarters, just north of Baraboo, Storhoff, 61, has led the reorganization of the predecessor organization — Wisconsin Dairies Cooperative — and formed Foremost Farms as of Jan. 1, 1995, through a consolidation with Golden Guernsey Dairy Cooperative of Milwaukee.

Wisconsin Dairies opened in 1963 with the merger of dairy cooperatives in Sauk City and Union Center. In just a few years the operation expanded with a series of major mergers. He also tested the concept of using the company's own increasing milk supply for its cheese-making operations.

In 1971, Storhoff came to work at Wisconsin Dairies and filled a number of key positions, including production manager and in late 1977, he was named general manager of the growing cooperative. He took over when the Dairies had $205.4 million in sales and $38 million in assets.

To show his executive ability, Storhoff became president of the Dairies in 1984 when sales hit $546 million and assets had grown to $144 million. Storhoff focused on the restructuring of smaller dairy plants by helping 16 separate organizations become part of Wisconsin Dairies.

In the Storhoff period, Wisconsin Dairies soon was recognized nationally as a Fortune 500 organization. The move to structure operations for the new century under the banner of Foremost Farms USA came in a consolidation move involving Wisconsin Dairies and Golden Guernsey Dairy cooperative, giving the new operation 5,300 members in Wisconsin, Minnesota, Iowa, and Illinois.

For Storhoff that was just a start. Within five months he brought together Foremost Farms directors and the Corporate Board of Associate Milk Producers Inc., to begin a feasibility study to see if AMPI's Morning Glory Farms Region could work out a plan in which Foremost Farms would acquire the Morning Glory Farms Region and its 3,200 member farms.

Earlier he was geared to add new income sources for Wisconsin Dairies members and the most notable was the purchase of McKesson's Foremost (note that name) whey processing system of three Wisconsin plants and its large research library.

Just recently Storhoff announced a major export sale of concentrate to Japan, a move expected to tap new markets and income for Foremost Farms.

It was the name of the three-plant McKesson operation which this year led to the new name of the cooperative — Foremost Farms USA.

These plants produce food ingredients for a variety of manufactured foods ranging from infant foods to bakery products — all value added products.

And, by December 1, the more dramatic move of Storhoff and Foremost Farms will be the expected final combination with the Morning Glory Farms Region of AMPI and its 14 processing plants and regional headquarters at Shawano.

At this point, Storhoff is already carrying out what he wants to see as the Foremost Farms plan for the new century.

It is all possible because of more than 7,000 interested and cooperative milk producers and a man named Don ***Storhoff***, who wanted to help them achieve a better dairy industry.

Selbrede Retires from Coopeative

March 11, 1990

This week there will be a retirement reception for Herbert Selbrede, chairman of the board of Wisconsin Dairies. He has completed 31 years representing dairy farmers and says he is "quite satisfied" with the way the cooperative is going.

Wisconsin Dairies is in the Fortune 500 list of top business organizations in the nation, with total annual revenues of $522 million.

Selbrede said while the cooperative has gained national ranking, the important measure is it "probably is one of the most financially sound firms on the list. That is most satisfying to Wisconsin Dairies member-owners."

Selbrede has been a dairy farmer for 42 years. That's one job he will continue with his son, Wayne, 37, so he can "keep in touch" with the family business. He and his wife, Berniece, started farming in 1947. They have four grown children.

The Selbredes have a 240-acre incorporated family farm operation called Leonval Inc. at Leon on Highway 27, south of Sparta in Monroe County.

But, who knows, even at 68, Selbrede may have more of a continuing farming career than he planned with the farm's 50-cow Brown Swiss herd: Wayne has opened a campaign for a seat in the Wisconsin Assembly.

In any event, the senior Selbrede, a soft-spoken, deep-thinking kind of cooperative dairy leader, has completed his compelling task as the top elected person of the 6,500-member cooperative.

He was the second chairman to hold the position, when he was elected to succeed retiring Melvin Sprecher of Sauk City.

Sprecher and Selbrede led the 1963 effort to form Wisconsin Dairies with the merger of dairy cooperatives at Union Center and Sauk City. Those cooperatives were the base for the dramatic growth of Wisconsin Dairies through a continuing series of mergers and acquisitions of plants in Wisconsin, Minnesota and Iowa.

Selbrede was secretary of the Dairies board until he was elected chairman in 1974. He became part of the system in 1958 when he moved his milk to Union Center to join a Grade A milk market.

Herbert Selbrede

He also was skilled in mergers, having helped cooperatives at Sparta, Tomah and West Salem consolidate their butter and powdered milk operations in the new Hiawatha Dairies at Sparta.

"These consolidations had to come because more processing equipment was needed to improve production efficiency. Such an investment had to be covered with a larger volume of milk and that volume was created through merger," he explained.

Selbrede was a devoted cooperative director and board officer. He never missed a board meeting from 1958 to 1990. He had plenty of meetings to attend. In state, regional and national posts he represented farmers as a director, officer or key committeeman like first vice president of National Milk Producers Federation, director and executive committeeman of Land O'Lakes Inc., director of Northland Foods and Wisconsin Federation of Cooperatives and its dairy division.

He regrets Wisconsin Dairies did not advance more with value added consumer products. It is a needed development, he said, so milk producers can get a higher return for their milk than the dominant bulk product market allows.

However, he is happy about Wisconsin Dairies' acquisition of Foremost Foods, a whey processing operation at three locations in Wisconsin.

"We were able to take this step because of better management and better financing," Selbrede said.

He's a Leader of the Herd

October 21, 1990

FITCHBURG — There must be an everyday thrill to be living on a corner lot where the intersecting street is named Lyman Lane, especially when it is named after the corner resident, Lyman D. McKee.

McKee, 79, was Wisconsin's Mr. Dairying for more than 30 years.

If you followed the production, promotion and marketing of milk and its products at home and abroad — McKee was a leader most of the time.

He was a national and international dairy celebrity, having spoken at meetings in every state but Maine and having sought new dairy markets in the 63 countries he visited to introduce new dairy products.

McKee was a dairy farmer, dairy cooperative president and manager, president of the American Dairy Association of Wisconsin and national president of the ADA, president of Dairy Society International and Wisconsin manager and overseas dairy market explorer for Associated Milk

Producers Inc.

Today he's retired and recovering from a heart attack and open-heart surgery. It has been a long recovery because a leg infection at the hospital slowed the healing process.

Despite the setback, he is just like the person he was when he was a leader in nearly every phase of the business when 152,000 dairy farms dotted the Wisconsin landscape. Today, the remaining 34,000 dairy farms are producing more milk with fewer cows.

He lives with his son Richard and daughter-in-law Linda at 2814 Jacquelyn Drive, at the intersection of Lyman Lane. The home is one of those on the McKee farms he and his brother Bill have developed since the late 1970s.

His brother died three years ago when they were developing 650 acres on four farms.

Half of the land is now devoted to homes, apartments, churches and businesses, south of the home farm on Highway PD or McKee Road.

But that doesn't mean McKee is away from farming. The McKee partnership has a milking herd of Holsteins and it rents cropland on 23 nearby farms for corn production.

McKee was responsible for farm operations while his brother organized the land development. Lyman's nephew, Bill McKee, has taken over the development work his father, Bill, started.

"I never thought we would come to the day when our farms would be developed as urban properties. Now I wish it was over with," McKee said.

The McKee brothers got into development business as a result of their grandfather's decision in 1848, the year Wisconsin beame a state.

That's when their grandfather, Dennis, came from Ireland and worked his way to today's Fitchburg area to establish his farm. His father, Edward, took over late in the last century and Lyman and Bill started farming together on Jan. 1, 1932.

It couldn't have been a more difficult time for their cow, sow and chicken operation, with the Great Depression under way and getting worse. But they succeeded even with milk prices as low as $1.65 per 100 pounds and hogs bringing 4 to 5 cents a pound.

Along the way they bought their first tractor with rubber tires — an Allis-Chalmers — for $700. They traded in their old steel-wheeled Wallace to finish the deal.

McKee served 12 years in the Naval Reserve. He was discharged in 1940 as a chief boatswain's mate three days before President Roosevelt mobilized Naval Reservists in the buildup of Naval forces prior to World War II.

Not only was the farm a base for McKee's dairy career, but so was the Madison Milk Producers Cooperative, which he served as president from the first day he was elected to the board of directors.

This oddity happened because his father, a director, did not like to drive to board meetings at night and for more than a year Lyman drove him to all board meetings and he sat through the long sessions.

When a vacancy opened, board members decided that since Lyman had been attending all the meetings, he should take the director's post and become president. He served as president for 20 years and when manager Harvey Thew retired, McKee was named cooperative manager.

Of all his accomplishments, and there were many, McKee said he was most proud of the retirement program he set up for members of the cooperative.

A news story got him interested in a retirement program because families found it difficult for two families to make a living on the same farm when a son or daughter took over the operation.

A retirement program could avoid the tough problem of expanding to gain a two-family income, McKee said, by letting the father retire, rather than living off the income of the farm operation.

A retirement fund was the answer, McKee said. By setting aside a few cents per 100 pounds of milk each month, a well-invested fund would provide enough money for the parents' retirement.

When McKee retired in 1976, the fund had grown to $4 million. But there was spectacular growth (with high interest earnings — 26 percent in one year) so today the fund totals $16 million.

Many of the retired farmers are able to live off an annuity of $450 a month, plus Social Security, and still not touch the principal.

Also, farmers's employees could join the fund and provide for their future. Four McKee employees are in the retirement program.

McKee also disclosed that it was a conversation with the late Harold V. Cobbs, a Lodi dairy farmer and Madison Milk Producers director, that got the initial studies under way to get dairy cooperatives working together to meet the changing milk market demands.

"We were walking to our cars after a meeting at the Park Motor Inn in Madison and Cobbs told how he was bothered because farm milk prices were $4 per 100 pounds when he started as a director and five years later they were still $4 per 100 pounds. Can anything be done to correct this?" McKee recalls Cobbs asking.

That conversation led to a meeting of all Wisconsin dairy cooperatives in 1964 after Glenn Pound, former UW-Madison agriculture dean, was asked by McKee for assistance to tell dairy cooperative leaders what they were doing wrong to improve farm milk prices.

That meeting opened the way for cooperative mergers, including formation of Associated Milk Producers Inc. AMPI now handles 12 percent of the nation's milk supply, McKee continued, and the nation's largest milk buyers are able to get the milk they need.

"I always regretted that Cobbs was never recognized for making the statement and asking the key question that changed Wisconsin's dairy industry, McKee said.

Williams' Dairy Program Puts Power in Hands of Producers

November 12, 1995

BARABOO — When more than 1,200 dairy farmers met in La Crosse eight weeks ago to discuss ways to improve farm milk prices, Robert J. Williams, Baraboo, a dairy consultant and dairy executive, heard their concerns.

Williams said he was impressed with the integrity and emotion of the crowd at the La Crosse Center as discussions showed that farmers believed they could team up and be effective in helping themselves gain a satisfactory milk price and hold it.

As he drove back to his home in Baraboo, Williams reviewed what farmers were seeking and what dairy programs there were to help them: A survey at the meeting showed their objectives focused on two key programs: gaining control of their farm milk prices and reinforcing their price objectives by being able to control milk production.

The simple truth of dairy success is to gain a price but not lose it by overproducing.

"These farmers were ready for a program in which they made the price and production decision themselves," Williams said.

So for the next three weeks he wrote a dairy program he calls PCP and PCP, an abbreviation which translates to "Producer Controlled Price and Producer Controlled Production."

He was encouraged after he introduced the plan to an Upper Midwest Dairy Farmers Committee and the chairman responded that the committee was considering his proposal.

Who is Robert J. Williams?

He is 67, raised on a family farm near Marshfield, was responsible for its operation for a year and then, upon graduation from Marshfield High School, was awarded three scholarships.

Williams majored in agricultural education and agricultural economics at UW-Madison.

The Korean War interrupted his doctoral studies, and he was commissioned as a lieutenant and assigned to an Army Transportation Corps unit in South Korea.

When he returned to Wisconsin he accepted a position as agricultural instructor at Lake Mills High School, where he not only taught the ag classes, but he was named wrestling coach.

He began working more closely with farmers after he was appointed office manager of the ASCS office in Jefferson County.

After that he joined the state Department of Agriculture as dairy marketing specialist.

As a dairy marketing expert he represented Wisconsin at Federal Milk Marketing Order hearings. He also developed the Wisconsin fluid milk report, which is still being carried weekly to track sales of fluid milk sold to regional and national markets.

After several years with the Ag Department he joined the new Wisconsin Dairies Cooperative. Williams held several Wisconsin Dairies offices, including membership director, public relations representative and chief of Grade A milk marketing.

Within four years he was named as general manager and chief executive officer of Wisconsin Dairies.

That lifetime of experience has led up to the possibility that his work with farmers eventually could involve the 125,000 dairy farmers in the nation.

Torgerson's the Name in Co-ops

June 28, 1992

WASHINGTON — Back in Wisconsin, the name Torgerson is prominent in the area of cooperative enterprise.

And in Washington, the focus on agricultural cooperatives also centers on the name Torgerson.

There is a connection.

Randall E. Torgerson, Washington, D.C., administrator of the Agricultural Cooperative Service in the U.S. Department of Agriculture, is the son of Truman Torgerson, Manitowoc, Wisconsin's man of cooperative organization, now retired, who made the business system successfully work for dairy farmers.

What's more, Randall's long career as administrator of the cooperative service traces back the days to the 1950s, when he accompanied his father to the organizational meetings of Lake to Lake Dairy Cooperative.

Truman Torgerson, as Manitowoc county agent, had been asked by dairy farmers to help them organize and be prepared for the changing dairy world. That led to the formation of Lake to Lake Dairy Cooperative at Manitowoc for dairy farmers living in the area from Lake Winnebago to Lake Michigan. The co-op now is a division of Land 'O Lakes Inc.

"I regard the opportunity to see Lake to Lake organize and successfully develop as my learning experience for the position I have held since 1975," Randall Torgerson said.

Wisconsin has had many USDA administrators over the years, but Torgerson's appointment in 1985 makes his the longest. He is 52. He is married to the former Susan Grundahl. The couple has two grown sons.

Randall Torgerson

Torgerson's move to Washington also got a start at the UW-Madison, where he received his masters and doctorate degrees in agricultural economics. He received an undergraduate degree at the University of Minnesota.

His first academic appointment was at the University of Missouri-Columbia, where he taught cooperative business, management and economics of cooperative action in agriculture.

Today he is recognized internationally as an educator, researcher and author on group action in agriculture. The focus of his agency is on preserving and improving the heart of American agriculture — the family farm. ***Torgerson*** has a ready answer for those who say Washington has forgotten the family farm, since this is the clear mandate given the Agricultural Cooperative Service, which Congress created in 1926.

This can be carried out because the cooperative form of business shows it can run with the best of all businesses, he continued.

The dispersed ownership means cooperatives have a major role in perserving and maintaining individual farmers in the nation.

Torgerson said the nation's 4,663 cooperatives, in the latest survey, showed a net business volume of $77.1 billion in fiscal 1990.

His analysis showed that the nation's dairy cooperatives dominate the agricultural business world with a 1990 net business volume of $20.7 billion. Grains and oilseeds rank second at $14 billion.

Dairy cooperative membership in Wisconsin that year was 29,096, highest total of all major dairy states. There are 45 dairy cooperatives with headquarters in Wisconsin.

New studies made by the cooperative service staff, Torgerson said, show that from the total farm marketings in the nation, cooperatives had a share of 27 percent.

It is Torgerson's belief now that cooperatives will find their business advantage in serving markets and membership by increasing value-added marketing through cooperative ownership of processing facilities. It is the processing of raw products, such as milk made into cheese or other dairy products that gives dairy cooperatives the value added advantage.

There are risks to such operations, he conceded, but "well-conceived and financed operations give (farm) producer-owners a fighting chance to continue as entrepreneurs in the ever-changing and increasing global food industry."

There were difficult years for cooperatives in the 1980s and the number of cooperatives fell as mergers and consolidations were used to maintain services to cooperative members.

Iowa, which was heavily organized with grain cooperatives, now is the state showing the greatest number of consolidations among local grain cooperatives.

But the financial condition of most cooperatives has improved and Torgerson reported them as "very good."

The Truly AmzingTales of Truman Torgerson

June 11, 1995

If you feel like getting away from it all for a few days, Truman Torgerson of Manitowoc, a retired dairy cooperative organization leader and former University of Wisconsin NCAA champion boxer, has written an exceptional book — his autobiography.

Its title is straight Torgy: "Digging up the Past," in 282 pages.

You'll just be amazed at Torgy's clear recall of life on the farm in the 1920s as the depression began and strangled the economy.

It opens with a story from the Torgerson farm in Dunn County.

It is a personal and compelling story. This is the kind of book a father and son would read together to gain new concepts of life and find new directions and inspiration that could lead to the "good life," which Torgy says he was able to achieve.

The book's strong themes tell of how important family life is, along with education (particularly the UW ag college), hard work (but having fun, too,) and the numerous friendships that Torgy gained and kept.

He was a graduate of the Dunn County School of Agriculture at Menomonie, which was accredited with a high school curriculum.

Torgy faithfully followed the recommendations of Randall Swanson, now of Madison, the ag school principal and agricultural instructor who became Torgy's confidante and counselor. Swanson advised the talented farm youth to get his education at the UW-Madison with a major in agricultural education and to train as a county agent.

He was county agent in Rusk County for a year and then was accepted as the Manitowoc County agent. His work with farmers flourished as he and leading dairy farmers worked to create and establish Lake to Lake Cooperative Dairy.

In high school Torgerson was known as Norsky; in his professional life he was known to friends as Torgy.

Retelling his years on the farm and his high school education, Torgerson fills the pages with interesting detail, but the UW years in the College of Agriculture and his experiences from 1935 to 1939, as a UW boxer, were exceptional reading.

He was a walk-on in both football and boxing, but his football days were short-lived because boxing was the popular sport and he was a good boxer from the start.

Things clicked for Torgy after he met UW Boxing Coach John Walsh — even though "my knees nearly buckled when I crawled through the ropes into the ring for the first time."

That first bout was with Dick Bardwell of Madison in the Contenders Tournament and he scored the quickest knockout of the evening, putting heavyweight Bardwell out in one minute and 5 seconds.

But the climax of Torgy's career came in his match with Americo Wozciesjes of Syracuse. Wozciesjes had a string of knockouts going and was the undefeated Eastern champion three years in a row when he met Torgerson in the Field House before a crowd of 15,000.

In the second round Torgerson smashed three hard rights to the Americo's face and suddenly Wozciesjes buckled. Torgerson got even stronger and won.

Fifty years later, the faithful still talk of that fight and Torgy's mastery. With the Americo fight Torgy won the NCAA national championship as a light heavyweight.

Dairy Farms Find a Friend

April 10, 1994

If you're looking for someone who keeps track of what 2,000 dairy farm families are thinking and doing, check with Richard La Croix, manager of one of three field regions of the Morning Glory Division of Associated Milk Producers Inc.

He'll tell you of the significant contribution one operating dairy farm can make to the local economy because most of its income is spent in the area where the farm is located.

Richard LaCroix

And he is talking big bucks — like $200,000 a year for a single dairy farm with a milking herd of 100 cows — something La Croix believes should be fully understood by farm lenders and the allied industries serving the dairy industry.

Even if 50 cows is more representative of average Wisconsin farms, that still means local expenditures in the range of $100,000 a year for each farm unit, he said.

La Croix is a member of one of 27 area councils across the state supporting the Dairy 2020 Initiative. The unit he is working with involves Dane, Green and Rock counties.

The 2020 program was organized a year ago. Despite the Legislature's recent failure to fund the program, there still is a push to continue the effort to serve family dairy farm needs and help farms adjust to remain or become profitable.

This 2020 program is a joint effort of dairy farmers, farm organizations, cooperative federations, state legislators, agribusiness and related farm business, state departments, UW and technical colleges systems, professional organizatioins and financial institutions.

The effectiveness and impact of the state's economic giant — the dairy industry — has great potential, La Croix said.

La Croix is hoping the 2020 program still has a chance to be funded by the state to help dairy farmers build a profitable enterprise for the new century. It is going to mean a new program to build needed milk production because Wisconsin dairy plants currently have only enough milk for 74 percent of full production.

All segments of the state's economy would benefit from income improvement and new jobs that come from economic successes of dairy agriculture.

La Croix, 49, of Stoughton has his headquarters in the AMPI cheese plant and office in Madison.

He was born and raised on a dairy-hog farm in Manitowoc County and is a former 20-year member of UW-Extension. He holds bachelor's and master's degrees from UW-Madison. He served as a county Extension agent in Calumet and Fond du Lac counties, and he worked with a private cheese plant before joining AMPI-Morning Glory Division, a dairy cooperative.

The challenge, he said, is to get the message to local communities so its citizens can appreciate the economic impact dairy farms and the dairy industry have on local economies.

He praised Kent Feeds for supporting 2020 educational sessions for Kent dealers to acquaint them with the details of the current issue: "What the Dairy Producer Means to Your Community."

Trading Enterprise Blocking State Dairy Farmers

March 3, 1996

If there is a goal shared by state dairy farmers in their quest for better markets, it is their desire to increase exports.

But in the rapidly changing world of dairy marketing, Wisconsin and the rest of the country are being blocked from potential markets by the use of a State Trading Enterprise (STE), Mario Castillo told area dairy farmers at the annual meeting of the Farmers Union Milk Marketing Cooperative of Madison.

Who is Mario Castillo? He lives in Washington, D.C., and is executive director of The Dairy Trade Coalition. The Farmers Union Milk Marketing Cooperative is a member and is represented by Stewart Huber, cooperative president.

Castillo has extensive experience with government agencies in Washington, including an assignment as chief of staff of the House Agriculture Committee. His notable achievement is gaining a victory in the cheese war with Hungary in 1992, when Hungary wanted to ship low cost Goya cheese into the United States to capture the U.S. market and replace American hard cheese producers.

Today, the power of a State Trading Enterprise, used by the New Zealand Dairy Board, to capture and dominate world market dairy sales is successfully blocking the U.S. and Wisconsin opportunity to compete in international sales.

How strong is that competition to block U.S. and state dairy sales in world markets?

Castillo calls the New Zealand STE "a monopoly," which the U.S. government thus far has refused to recognize even though it has a chance to do so through legal remedies specified under the General Agreement on Tariffs and Trade.

He told the farmers that millions of dollars are at stake in this trade dispute, which can be corrected through action of four federal government agencies — the U.S. Department of Agriculture, the Justice Department, the Internal Revenue Service and federal law enforcement agencies. Castillo said all it would take is an order for investigation and enforcement of GATT rules from Congress.

Such a dairy market opportunity, he continued, is the key to a higher dairy farm income. When those potential international market gains are spread through farm milk checks, they would be sufficient to keep small family-owned dairy farms, like those with 50 or so cows, in business.

Ever since GATT was endorsed by the U.S. government, Castillo has been organizing an attack on the STEs and now is gaining support in Congress and from major European government interests.

"At this time it is unclear whether or not the (Clinton) Administration will heed our recommendations. But I can tell you that representatives from two of the presidential candidates have expressed an interest in our work," Castillo said, without naming the candidates.

He had high praise for U.S. Sen. Herb Kohl, and U.S. Reps. Steve Gunderson and Dave Obey for their leadership on dairy export issues, including ways to solve the State Trading Enterprise problem.

Consultant Assists Hungarians

December 22, 1991

For nearly 40 years, Charles Farr has been engaged in marketing with agricultural cooperative organizations that work to give dairy farmers a voice in public affairs.

Now he is taking the cooperative message internationally through a program called Volunteers in Overseas Cooperative Assistance (VOCA), a program that's funded by the U.S. Agency for International Development (AID).

The first place to get the Farr assistance was Hungary, which is going through Eastern Europe's privatization efforts to form a free market economy since the collapse of communism.

However, some of the past lingers and in Hungary the large state farms — a carryover from the communist system — with 600 cow dairies and equipped with modern Swedish milking parlors and cow housing systems, are going to stay in operation, Farr said.

This will happen, despite a move to return 15- to 20-acre parcels to families that lost land when the communist-style farms were formed.

Farr saw all phases of Hungarian dairying, including its milk marketing and promotion programs. Such short overseas tours by marketing and distribution experts, primarily in food related fields, are being sought by the U.S. government to help speed development of free economies in Eastern Europe and the new commonwealth, which is taking the place of the Soviet Union.

Farr's tour lasted two weeks and he spent most of his time interviewing officials who will be among the people creating dairy marketing systems.

In recent years, the Madison man has been a consultant

to the Wisconsin Milk Marketing Board and to Will Dahl, the board's chief operating officer, after completing his career as director of the dairy division of the Wisconsin Federation of Cooperatives.

Farr, who holds two degrees in agricultural economics from the University of Nebraska, came to Madison in 1964 after 11 years as staff economist with Twin City Milk Producers Association in St. Paul.

Charles Farr

His assignment in Hungary came at the request of the country's National Federation of Agricultural Cooperatives and Producers, called MOSZ, which is in charge of marketing all commodities and maintains financial, fertilizer and chemical services in all 19 counties. It also has a research function.

Farr said the country of 10.6 million people is about the size of Indiana. Its 25,000 dairy farms (some with herds of only two to five cows) produce 6 billion pounds of milk a year, which is one-fourth of Wisconsin's annual production.

More than half of the milk is used in manufactured dairy products, primarily cheese. "About 55 percent of the land rates as good farmland," he said, "and average yields of barley and wheat are higher than those in the U.S."

But Farr said the overall dairy system is in chaos and competitive forces will dictate the success of a new marketing system and improving long-run returns to the dairy industry. A change is needed, he said, because processing plants have a monopoly and have authority to reject shipments of milk from farmers.

The principal dairy goal now is to learn how to make up a marketing system — the process they call privatization. Most people asked what the U.S. dairy system was like, but Farr only used parts of the U.S. system to suggest solutions so officials could begin discussions leading up to establishing a system of their own.

He recommended a system that will involve government money, classification of milk according to the way it is used and new regulations making it possible for farmers to have a say in operation of the marketing plan.

Farr reminded his hosts that dairy farmers in Hungary had the largest investment of any group involved in the dairy industry, including processing plants.

Budapest, the capital, has two McDonald's restaurants and a Burger King. Dairy officials watch their advertising with interest and think they should also advertise their products.

Farmers say McDonald's has created a market for their cows culled from dairy herds.

Farr took pictures of "an absolutely beautiful supermarket" and American brands of cooking oils, catsup, candy bars and the two colas, Coke and Pepsi.

In his final report he pushed for a system that would improve the quality of milk to start improving consumption of dairy products. He said the milk price has increased 40 percent in the past year and triggered the 20 percent drop in dairy sales.

There is a surplus of milk and its dairy products could move across the border of the old Soviet Union, but the dairy export market is lost, mainly because the Soviets lack hard currency, Farr was told.

Hungary's government used to have a majority of representatives from agriculture, but now only a single farmer has a position of influence over food matters.

Business Thrives in Pennsylvania

March 6, 1994

MANHEIM, PA. — In traveling around with Wisconsin dairy farmers, you soon learn what farm visitors want most. They want to find a farmer with ideas and a record of successes as an entrepreneur.

A busload of Wisconsin dairy cooperative leaders in the area for a dairy conference found their farm and business entrepreneur in Lancaster County, Pa.

This county, with an excellent lay of land between the mountains, is one of the counties that continues to have a better farm financial record than even Dane County, Wisconsin, which to many is still a Garden of Eden.

Their generous host for an afternoon and evening dinner was Noah Kreider, a dairy farmer who was plugging along like many other dairy farmers early in his farm career.

Things changed for Kreider, however, when his parents bought the farm of his mother's family. This transfer took place right in the middle of the 1930s Depression.

Noah describes Kreider farm events rapidly as farm after farm was available in the early years of World War II and after the war.

"And since we had nothing to lose, we bought each one as they became available," Noah said.

Over the years, Kreider and his brother, Richard, bought 25 farms and the lands of 12 of them are joined together. When the bus was on a ridge road he urged his fellow farmers to look over the easy sloping land because what they saw was all joined together as Kreider Farms.

Today, Kreider Farms tills 3,000 acres and when the four big Steigers get purring across the fields it is in sharp contrast to the Amish doing the work with four and six-horse hitches.

Noah's brother was killed in an accident, but Noah carried on and the Kreider enterprise expanded. Today Noah is in a partnership with his two sons, who earlier told their father, "buy the farm and we'll stick by you."

Today the partnership prospers with excellent practices in all business lines.

"The Lord has been good to us. What we have is the Lord's, the bank's and ours," Noah said with a prideful smile.

Noah has a philosophy that people want good food and service and as they make their demands known, he provides them with their needs.

Along with the farms, there is a growing business establishment and each one is sparkling clean — a mark of distinction that identifies a Kreider business.

Kreider Dairy Farms is still a comforting and reliable name to customers, even though it has increased to a herd of 900 cows.

The cows are milked in parlors where the milking goes on 24 hours a day. With the young stock, he has 1,100 dairy animals on two farms.

The Kreiders are considering a dairy herd expansion because the kosher milk operation they built up to 200 cows is not a big enough base for the demand from Philadelphia to New York.

To get the kosher label, a Jewish Rabbi lives comfortably near the kosher milking and packaging operations. The rabbi offers a prayer after each milking and at the milk processing plant.

But there is more — much more than the dairy farms. In 1977, Noah started laying hen operations on many of the farms he acquired.

By 1982 he had millions of layers to serve major markets from Baltimore to New York.

In 1987, he started new laying operations in the Harrisburg area, a considerable distance north of Lancaster County.

One of his sons is in charge of all the poultry operations.

About 30 percent of the milk produced is used for whole milk and ice cream at the Kreider restaurants. The surplus goes to an Atlantic Dairy Cooperative plant.

When a farmer on the bus asked Noah what contributed to his success in the farm business, he said that was an easy question.

"We just made the right decisions and that, of course, came with the experience we gained," Noah added. When the Kreider planners talked of building an ice cream shop in 1979, they took a close look at the costs and determined quickly that it was better to build a full restaurant.

"We made that decision even through the highest death rate among businesses around here was restaurants." That wasn't the prospect now, even though the Kreider restaurants face the most competition of any of their operations. The tour group went to one of the eateries for dinner on Noah. The building was 240 feet long, not including a banquet room.

Noah's youngest son, 29, is in charge of the restaurants and the stores providing the outlet for Kreider milk and ice cream products as well as food.

There is a 450-acre irrigated potato operation to provide the spuds for potato chips.

Each month the Kreider payroll is divided among 450 employees, including those who work in the firm's four restaurants.

Among the statistics he dropped was one on his monthly electric bill at $86,000, "even if we have the best rates around."

As his sons gain experience, Kreider said he is trying

to step back from the operation, "because that is the way it has to be."

He was also asked if he would do it all again and build this kind of enterprise.

"Let me think for a minute," Noah said. And later, he answered: "Yes, I would do it again. I really enjoy it."

Blue Mounds Woman became a Barneveld Legend

December 30, 1990

BARNEVELD — Betsy Thronson may already be a Barneveld legend.

A dairy farm wife of rural Blue Mounds, she is the organizing genius who stepped in to feed the multitude of community residents and the volunteers who provided help and hope after this Iowa County village was ravaged by a devastating tornado just past midnight on June 8, 1984.

This terror of nature changed Betsy's life.

A lot of volunteers and townspeople worked to clear away the wreckage, but the focus was on Betsy. She was in Appleton at a meeting when the winds struck the Iowa County community, but within hours she was in Barneveld making an assessment of what was needed most and what it would take to fill that need — mainly a place to feed the people and to provide a base where people of the community could come together to support each other.

Betsy is that kind of person, primarily because she always has had a personal desire to help, gained from her parents and grandparents in the Des Moines area of Iowa, where she was born and educated.

"My philosophy of life is in my heritage because my relatives were people who felt it was their Christian responsibility to do those things that could help others," she related in a chat this month.

Betsy told how she always had that feeling, but only at a family reunion last summer was she able to trace that family trait of helping others to the teachings of a segment of the Lutheran church.

It was her grandma and grandpa, she said, who always had their hearts and hands out to help others. "That's the way I live my life, at least I hope I do," she added.

"I don't like to talk about it , because what you do for others should not bring any recognition. You just hope that those you helped also will help someone else in need. That's how it should be and why the tradition of service can be passed on to more and more people," Betsy added.

When she talks of her youth she quickly reminds you she is "a flatlander from Des Moines" — born Elizabeth Foshi. It is a French name from the bayous of Louisiana. But Betsy is an American, "really — my mother was Norwegian and my dad was French and Irish — a real combination, stubborn Norwegian with the humor of the Irish and sexy like the French," she laughed.

Not only that but she has a contagious smile and lilting laugh.

But there also was a time when Betsy was called "Elizabeth" by a proper woman of society on the Chicago Gold Coast. Betsy worked there six months and was in charge of the front part of the parlor, serving meals and "polishing the silver service every single day."

Betsy took the "position" when she suddenly had to leave her studies at Wartburg College when her father was stricken with a heart attack. She took the Chicago job for six months to pay off her college costs. She also worked in Des Moines for Meredith Publications.

And how did she get to Wisconsin?

Thank Glenn Thronson, the dashing earth-mover operator who was working in Iowa on a road project when he met Betsy and married her in 1959.

The marriage also changed her life. They became dairy farmers after the couple decided the best place to raise a family was on a farm. That's where they raised five children, two sons and three daughters.

"Glenn brought me back into the hills of Iowa County and I couldn't find my way out," she laughed.

"It was a major decision, especially for me since I wasn't raised on a farm and didn't know how I would react to life on a farm and face all the responsibilities of dairying. But you make a commitment and live with it. That's the way it has to be," Betsy said.

So she organized and became a leader in dairy and agricultural organizations, especially for farm women. She started out in the old Pure Milk Association and became more and more involved in dairy policy as major cooperative mergers of the late 1960s changed the power position of farmers.

"I was raised in a union family in Iowa and always believed there was strength in unity. If members worked collectively for a goal, their reward could be attained.

"Cooperatives and farm organizations are only as strong as their memberships. For members it takes loyalty and work to make cooperatives successful," she said.

As for Betsy, the more she became involved in dairy cooperative affairs the more she felt a personal dedication to working to help fellow dairy men and women. "You have to have that dedication," she insisted.

And look at her record in the next 11 years: She organized one of the first chapters of the new Wisconsin Women for Agriculture. Then she organized the first group of women in Associated Milk Producers, then organized AMPI women in the state and served as president for four years, also organized women in five states of the Mid-States Region of AMPI, serving as president for four years and then received the highest calling of her career, when AMPI members elected her as an AMPI district director, a job she held for three years.

She still insists that problems dairy farmers experience today wouldn't exist if cooperatives joined forces for the benefit of their members.

Her recommendation is that directors of cooperatives as well as local school boards, state and federal representatives all serve a limited number of terms.

When the tornado swept through Barneveld, Betsy had new organizational challenges.

She quickly provided temporary feeding facilities and fed the volunteers who came to Barneveld to help the people restore their homes and businesses.

She made plans to move into the Brigham Town Garage by the night of the first day of Barneveld's revival so the kitchen could be ready one day after the storm battered the village.

The garage was the only large, safe building in the village that could handle the kitchen and the scores of volunteer workers and residents.

Then Barneveld residents took refrigerators and freezers to the garage kitchen, not only for her use but also to get electric power so their stored foods would keep.

"We brought in large ovens from the school to keep pace with the meal demand."

"Betsy's Kitchen" was born. Ann Luthold, a Barneveld senior citizen, first suggested "this is Betsy's Kitchen," and Betsy still has the magnetic sign that made it official.

The garage was used for the kitchen and serving area until Sept. 28, more than three months after the tornado hit.

Betsy said the kitchen served 1,000 people a day at first. Then the count settled down to 200 a day, Betsy recalled. Work days for Betsy ran from 6 a.m. to 10 p.m. She stayed with the kitchen for 111 days.

In addition to a place to eat Barneveld needed a place

where residents could meet and support one another. Betsy provided that service, too.

Scores of volunteers depended on Betsy. She has a book with all the names of the volunteers.

The first day Betsy started her feeding effort she got cheese trays and milk from the Pure Milk Association plant in Mount Horeb, contacted are churches to prepare sandwiches, brought in coffee and set up places for people to eat, using round cattle feeders to hold "table" planks.

The Red Cross provided food funding and later contributions made food purchases possible. "We were able to serve any type of meal a person could get at home," Betsy said.

There were brighteners for Betsy, too. She recalls the little Mennonite lady, struggling to slice a ham and saying: "Oh, I wish I had my electric knife."

In the fall, with the garage kitchen closed, there was talk of having someplace for the people of Barneveld to eat in a family setting.

In the spring of 1985, Betsy and a group of silent partners agreed that Barneveld would have a family restaurant, "Betsy's Kitchen."

"We mortgaged our farm and our souls after a long decision period. But we opened on Feb. 28, 1987. To me it is still unbelievable," she continued.

She got help from the UW-Madison small business center for the feasibility study and the Southwest Technical School at Fennimore to train her staff.

The restaurant is going strong today along Highway 18 and 151 at the entrance to Barneveld. It is there as the result of good advice, a lot of help and good support from the public, she said.

"I only hope a recession doesn't hurt us," she concluded.

Bird Builds Career Serving Dairy Farmers

August 29, 1993

BROOKFIELD — Jim Bird, 42, general manager of Milwaukee Cooperative Milk Producers, moved into the leadership ranks of the dairy industry by knowing farm families first.

He grew up on a dairy farm near Brownsville, where his father, Robert, and grandfather farmed together.

Jim Bird

After graduation from Goodrich High School in Fond du Lac, he continued his education at both the UW-Fond du Lac campus and at Moraine Technical College.

This year, he will celebrate the 21st anniversary of his marriage to his high school sweetheart, Deborah Huck. The couple has two children, Benitt, 16, and Baritt, 12.

Bird built his entire career serving dairy farmers. He started with the Dairy Herd Improvement Association in Fond du Lac County, then joined Milwaukee Cooperative Milk Producers as a fieldman 17 years ago.

Bird's territory for the rapidly growing cooperative was most of southern Wisconsin. When he started, the cooperative had about 550 members. It now serves 2,200 dairy farmers in 35 counties.

When the milk cooperative was organized in 1916, membership was concentrated in the areas close to Milwaukee. But as the Milwaukee metropolitan area grew, the number of dairy farms there was reduced.

Milwaukee Milk expanded by signing members in southwest and east central Wisconsin. Now, the cooperative's membership center is Madison, and that's where the group holds its annual meeting.

Bird was named general manager seven years ago, providing leadership for the organization that ranks as the 21st largest cooperative in the nation. Its members ship 1.74 billion pounds of milk a year.

As the cooperative's membership and milk production have expanded, the organization has put emphasis on representing members shipping milk to proprietary plants.

"Furthermore, our directors were concerned about health care costs and health insurance," Bird said. "The cooperative was one of the first to sign its entire membership with WPS health insurance."

Convincing Washington lawmakers to change farm policies is a time-consuming job, but the cooperative is represented on dairy policy by the National Milk Producers Federation. Bird is Milwaukee Milk's representative to that organization.

Conditions on farms are changing rapidly and one of the major tasks in Washington is to acquaint members of Congress from urban areas with the emerging dairy situation.

On dairy farms in Wisconsin, Bird said, the time has come for producers to recognize the need for more specialization and division of labor. In larger operations, the farmer-owner no longer will take charge of the crops, milking and other farm chores, but must concentrate on those operations that will improve income, he said.

Dairy Exec Find a Home in California

June 21, 1992

WASHINGTON — About a year ago, James Tillison was living in Madison and working in Jefferson with the Morgan Myers advertising agency.

Since then he has set up an office in Sacramento, Calif., where he is executive director and chief executive officer of The Alliance of Western Milk Producers — and the only employee.

Seven dairy cooperatives and two producer groups make up the membership.

Once he was an executive director for the Wisconsin Cheese Makers Association.

But he kept getting offers that were difficult to turn down and that led the way to Jefferson and Sacramento. That meant some pretty good career choices for a person who came to Madison in the first place when he was 35 and took over his father-in-law's insurance business.

Earlier he was a student at the UW-Madison and UW-Milwaukee in business administration, another of his skills he put into practice at the Waukesha Motor Co.

His insurance agency was a specialized group focused on coverage for cheese plant owners. His clients also were real-life business teachers and the chats about the business soon made Tillison a specialist in Wisconsin's cheese industry during some of its most dramatic growth periods.

But this profile of the 46-year-old executive was written in Washington, where he spent five days sitting, reviewing and talking to the policy pros of the dairy industry attending a federal milk marketing hearing.

Tillison didn't testify, but said he probably would file a brief in two months.

He will comment on the 30 hours of testimony and cross-examination on proposals to develop a new basic formula price so Wisconsin and Minnesota can continue to set the basic price of milk for the nation.

Right now he works for California dairy farmers. (Whoops, they don't like that name, and Tillison said quickly, "they want to be known as dairymen because they make their living milking cows.") Also, their operations are not dairy farms, but dairies — even with only about 2,500 such operations in the whole state. That's less than Wisconsin may have in two adjoining counties in the north central part of the state.

Many in Wisconsin ask him about California surpassing Wisconsin as the nation's top milk producer, but Tillison quickly cuts off that kind of talk.

"If we do produce more than Wisconsin, we aren't going to change the license plates," he said, adding that "we're not talking quantity of milk, but we are talking about its profitability."

That word "profitability" is one a lot of Wisconsin dairy farmers don't want to hear used.

And what about the California drought? "The perspective of the drought is greatly overblown back East. I don't think it has had any impact on feed prices, which are very favorable. The drought is not something I heard dairymen complain about," Tillison explained.

If Tillison is to take another step up on the dairy ladder, it probably will be nationally in the realm of getting consensus among dairy groups.

He is starting to get some consensus on dairy policy and said California used its consensus during debate on the 1991 dairy bill in Washington last year to seek a compromise over the milk price support level.

It didn't succeed, but there was support for the move,

and "that's a start," Tillison said.

He likes California and he and his wife, Jullie, have a Wisconsin connection — their daughter, who is a UW-Madison student wanting to be a teacher. "She stayed in Wisconsin and mom and dad left," he said.

Shortly after moving to Sacramento, Jullie began to think the weather was always the same and longed for Wisconsin's changing patterns.

In October, she flew to Madison to visit her daughter and survived the temperatures in the 40s and some rains, but told her husband the Sacramento weather wasn't so bad after all.

She also learned that annuals she likes in her flower garden bloomed as expected the first year, but growth and blossoms from the same plants came unexpectedly in the second year.

Tillison likes the weather, too, because the grass doesn't grow fast, "and that's just great."

Chapter 3

Cheese Makers

Cheese Makers' Chief is Off and Running

October 28, 1990

Paul Julius stepped into his new job as executive director of both the Wisconsin Cheese Makers Association and the U.S. Cheese Makers Association at a most challenging time for the state's $10-billion industry.

Julius, 40, who has 14 years experience in trade association management, took office just as cheese prices fell sharply at the time of year they normally increase.

His appointment last month to head state and national cheese makers' groups in Madison, started when cheese makers were facing increasing inventories that touched off the price drop, when they had to protect their interests in the final version of the 1991 farm bill and when it was necessary to prepare for a nationwide hearing on the future of federal milk marketing orders.

"You ask how I could acclimate myself in such major business? All I can say is, 'just jump in with both feet and get acclimated very soon,' that's what," he said.

"Challenging times," Julius added with a slight smile.

He is soft-spoken with such a calm manner you know instantly he handles problems in the same way. And, being a distance runner who tries to get in a few miles every day — although his schedule makes it more and more difficult — Julius was in good physical shape to get going at his Downtown Madison headquarters in the Tenney Building.

When he has the time, he said, he still likes to golf, fish "and walk around with a canoe on my head in northern Wisconsin."

Julius grew up in Pittsburgh, where his father was a

stamping mill worker and his mother a clerk in a store. He worked a year after high school and then registered for Penn State University courses, first at a Pittsburgh campus and then at State College, Pa., where he majored in English. His first job was five years as an English teacher in Rochester, N.Y.

He is married to Helen, whom he met at the university. They have two sons, one in high school and the other at UW-Madison. Helen will join the teaching staff of Madison Area Technical College.

In his last association management position he was with five others in a management team, managing 27 trade associations. Each member handled about four associations.

That was in Chicago where he moved from Rochester, N.Y., after serving as training director with a retail lumberman's group. He moved to Chicago to work with the National Sash and Door Jobbers Association and, later, as executive vice president of the Chicago Lumber Trade Association.

This all changed when he read an ad in the paper and became a name in the list of 125 persons applying for the Wisconsin position.

To give you some idea of what the state cheese industry is like, Julius said about 80 percent of Wisconsin's annual milk production of 24 billion pounds is made into cheese.

The cheese makers' proposals for the milk order hearings insist that they reflect Wisconsin's and the Upper Midwest's relationship to other sections of the country and world.

"There is competition coming from overseas and we are in a global market — a global economy. In two years we will be looking at Europe as one market and that is going to be a tremendous challenge, not only for the dairy industry, but lots of other industries in this country," he said.

He noted that in the World Championship Cheese Contest judged earlier this month at Green Bay, 16 classes of cheese were judged. Seven of the top award-winners were from the United States and five of those from Wisconsin. The remaining nine honors went to cheese makers in other parts of the world.

"That means there is a lot of good cheese in the world," he continued. The world champion cheese was an Emmentaler variety entered by Josef Schroll of Austria.

Julius is a stickler for emphasizing high quality milk from farms and high quality cheese from the plants, which will be essential if the United States is going to compete for its share of domestic and international markets.

Without quality production we would be losing a market opportunity, he said.

It is up to the industry to set high quality standards, he added.

Cheese prices in the U.S. generally follow the trading on the National Cheese Exchange at Green Bay, where the market is open for a half hour each Friday morning.

Julius said it is difficult for the Cheese Exchange to control the prices of cheese on a nationwide basis.

"I think the market is far bigger than the Cheese Exchange. I don't think cheese makers should feel intimidated by what happens there (at the exchange)," he said.

The hearings to determine federal order regulations and milk pricing "hopefully will result in a workable plan for the industry and cheese makers.

"No one wants to see continued consolidation of dairy plants. We don't want to see fewer cheese plants or fewer dairy farms. We want policies that will enable milk producers and milk processors to grow. Those plants and farms represent a way of life for many people," Julius said.

Father, Son Lead Cheese Making Family

November 10, 1991

GREEN BAY — For the father and son Baker team from St. Cloud in Fond du Lac County this is the year to celebrate Wisconsin cheese making and their successful string cheese business, Baker Cheese.

First, Francis Baker, 68, and his son, Richard, 47, are the first father and son team to have each served terms as president of the Wisconsin Cheese Makers Association.

The elder Baker, served as head of the state association in 1976 and, Richard, the association's new chief was elected this year.

"And what a year to be president," Richard Baker said, "since this is the 100th year of the state cheese makers organization."

From dairy farms on to cheese consumers nationwide — and with an additional slice for exports — Wisconsin's cheese industry leads the nation with $10 billion a year in business.

What's more, Wisconsin cheese makers for years have been an unseen power because they establish the basic milk price for dairy farmers in most of the nation.

The Baker family cheese firm, like other state cheese makers, also is celebrating an anniversary this year — its 75th year since Frank Baker, Francis' father, bought the plant in 1916.

Although the plant has a St. Cloud address, it is actually six miles south of St. Cloud, at the intersection of Highways G and T, a short distance south on G from Highway 23.

Even though it is "out here in nowhere," as Richard's wife, Karen, says, Baker Cheese is big business. The plant depends on about 100 dairy farmers for quality milk. There are 65 employees divided into three shifts to meet the demand for string cheese from consumers in 32 states, according to Richard Baker.

Richard Baker is president and also serves as financial officer and sales administrator for the company. He said 70 percent of the plant's production is string cheese, a specialty operation the Bakers started in 1978 and have expanded since.

The Bakers are proud of their four-generation business. "And, we can have a family reunion every day," he chuckled.

"My father, who retired in 1989 (or about that time) still gets to the plant nearly every day, I am in the office; my brother, Jack, in charge of purchasing, shipping, receiving and quality control; another brother, Bob, handles the cheese production area; and my son, Mark, 21, is our utility man getting acquainted with all phases of the operation during his shift. Starting at 11:30 a.m., my wife, Karen, handles the office and my sisters, Mary and Diane, run the cheese store, open daily, 9 to 5, except on Sunday and after 1 p.m. on Saturday," Richard Baker said.

He said the plant is run on a strict business principle: "We don't make a pound of cheese until we have an order to fill. We've been operating like that for the past 20 years or so."

It means taking some risks because they have milk coming in each day.

"But when we have extra milk, we sell it, which means on some days we lose," he added.

When the string cheese operation started in 1978, Francis Baker predicted a short time later that "some day this is going to be our major product. His prediction has come true," Richard Baker said.

One of the reasons string cheese has become such a favorite is that it has what he calls "consumer-plus features."

The mozzarella cheese is in convient packages, with the most popular one holding an individual, one ounce 1/2-inch by 5-inch piece of cheese, primarily used in lunches for school children, but now also being used in lunches packed for adults.

Furthermore, people perceive it as good nutrition and a reduced fat item, Baker continued.

The Bakers also contributed to the reputation of their cheese, by consistently winning best-of-class honors in cheese contests, like the recent U.S. Championship Cheese event.

Richard Baker also sides with the views of many dairy farmers who believe that with a changing dairy market there has to be some national effort to control milk supply. He said he believes that a supply management program compromise will be passed by Congress and a voluntary diversion plan will be initiated for the benefit of dairy farmers and the industry.

A Vision that Says Cheese

April 4, 1993

GREEN BAY — Wisconsin produces about 2 billion pounds of cheese each year and today there is a ready market for all of it.

But Monroe cheese maker Randy Krahenbuhl has a vision for a production and marketing system that could improve the returns for both cheese plants and milk producers.

Krahenbuhl, 38, president of the Wisconsin Cheese Makers Association, and his wife, Michelle, took time out from the most recent Cheese Industry Symposium and their active schedule to talk cheese and where the industry, especially the small plants, is going.

When Krahenbuhl was a junior in Brodhead High School, his father bought the cheese plant at Monticello — the Town Hall Dairy Cooperative — and Randy became part of the family workforce.

Michelle and Randy met in high school. Michelle's parents were owners of the Chalet Restaurant.

The couple were married in 1972 and live in Monroe. They have three children, Jamie, 17, Amanda, 15, and Emily, 11. When Randy graduated from high school in 1972, he decided to train for a career in auto mechanics. He took courses at Madison Area Technical College to get the needed training.

He did get a job after his two years of school. After a year as a mechanic at Dearth Motors, he found that being a mechanic wasn't his calling.

So it was back to the vat in Monticello where he used his cheese making skills to turn out Swiss cheese blocks and a few Emmenthaler wheels.

The cheese business carried the family for a lot of years — such as in 1989 when there was spectacular demand for cheese because of milk shortages in competitive markets.

This operation continued until recently when his father, Walter, reached 65 and was thinking about retirement. That meant the plant was up for sale.

Also the farmers wanted to sell their part of the plant, since there were just three stockholders left. The family wanted Randy to buy the operation, but he thought of the financial risks and took his time making "a hard decision."

But the situation changed suddenly when Swiss Valley Farms of Davenport made an offer to buy the plant, making his decision unnecessary.

Today, Randy is plant manager and has the time to combine cheese making skills with his dreams for a new approach to cheese marketing.

The Swiss Valley purchase gives him time to plan and

he is constantly developing new cheese specialties because he believes that's the way small cheese plants can survive in competition with the large plants.

He also has used his time to make exceptional cheese and he has the reminders of high honors he won at the Wisconsin State Fair and the Governor's Sweepstakes competition, the American Cheese Society, which met at UW-Madison last summer, and in the U.S. Cheese contest.

Ever since Randy started in the cheese business his goal has been to save time and increase cheese production efficiency.

He is encouraged over the plant efficiency and cheese marketing experiences and he travels to see how foreign cheese competitors gain markets in the United States.

"It is not just the little guy, but everyone in this business is in competition. What I learned is that we have to be more realistic in approching the marketing phase of the business," he said.

Buholzers Carries on Family Tradition

September 4, 1994

There is a beautiful story about family business coming out of a cheese factory in Green County with the election of Ron Buholzer as president of the Wisconsin Cheese Makers Association.

Buholzer, 47, carries on a strong Swiss tradition as a cheesemaker and, now, as the person with overall responsibility for operations at Klondike Cheese, midway between Monroe and Argyle along Highway 81.

Buholzer started his term last week at an elaborate reception for Mexican food authorities who wanted a taste of Wisconsin cheese for possible marketing in Mexico.

The visitors got the best of show in a 45-foot-long display of cheeses at the Concourse Hotel in Madison. It was arranged by the Wisconsin Department of Agriculture, the Wisconsin Milk Marketing Board and the National Dairy Promotion Board.

The Mexican market offers the cheesemakers an interesting opportunity: The middle class is the fastest growing sector of the economy.

Ron Buholzer

Buholzer has confidence in the $10 billion-a-year Wisconsin cheese industry.

"We are going to continue to gain in milk and cheese production with a lot of dedicated people who want to stay in farming and in the cheese manufacturing business," he said. "They all are making changes to serve the consuming public with the highest quality milk and cheese. All those I know in this business have the same attitude.

"I just hope, in the short term, the introduction of international trade is not too much at the expense of our dairy farmers."

This is a worry among cheesemakers because the U.S. GATT (General Agreements on Tariffs and Trade talks) negotiators have granted all 12 European nations and 15 other countries permission to ship imports of 30.9 million metric tons of major varieties of cheese annually — including American cheese — for six years.

The historic Klondike factory was started by a small group of Green County farmers as a cooperative in the late 1800s. The farmers built a plant and hired a cheesemaker who brought equipment and began production of Swiss cheese in those big cheese wheels, weighing from 180 to 220 pounds each.

Buholzer's grandfather was hired as cheesemaker in

1925 and Buholzer's father, Alvin, took over the position in 1947 after completing service in World War II.

"My father taught me all about making cheese, and we continued producing the big Swiss wheels until January 1970," Buholzer said.

Klondike then switched to other varieties of cheese that were more in demand. It first shifted to Cheddar with the help of Ron's two favorite cheese experts, Bud Meister and Jerry Schell, both with the Borden Company's Cheese Assembly operation at Richland Center.

Klondike later moved into production of longhorns, mozzarella, and provolone cheeses.

It now sells a feta cheese, Odyssey, a Greek specialty cheese. It also produces Muenster cheese, a needed sales companion to feta because a speciality cheese alone cannot provide the needed sales volume.

Buholzer named about a dozen states — from New York to Florida and to Washington state — where Klondike's feta cheese is distributed. And Muenster "is a bright spot for us, too," he said.

Klondike has contracted with Dan Carter, Mayville, the ranking authority in Wisconsin in specialty cheese marketing. Buholzer said Carter has developed the successful marketing plan for Klondike.

Klondike is a family corporation, known as A.X. Buholzer and Sons Inc. The stock is held by Ron and his brothers, Dave and Steve, and his mother and father. The senior Buholzers have retired, and each of the sons has assignments in the plant or in field operations.

They get the milk for the plant from 65 dairy farmers. Two of the producer families have ties to the original farmers who formed the cooperative plant in the late 1800s.

The producer families will meet for their annual picnic and fun day this month at a Green County park.

Monroe Man Rises to Top in Cheese Making

September 3, 1995

MONROE — Wisconsin still ranks as No. 1 in cheese production with more than 2 billion pounds a year — for which it takes the skill, judgment and energy of a large corps of cheesemakers.

Each year one of that number is selected as the Grandmaster Cheesemaker in the Governor's Sweepstakes Cheese competition.

This annual public event is part of the Wisconsin State Fair and is another step taken by Wisconsin's $10 billion cheese industry to perfect the taste, acceptance and reputation of Wisconsin cheese.

For instance, the cheese contest drew 167 cheese entries this year and 13 cheesemakers won class honors. Bruce Workman, with one of the Gruyere (pronounced Grie-yehr) cheeses made at Roth Kase won the top grandmaster award.

To acquaint people with the way the cheese business operates, he said, "The award is not mine, but is the award of everyone in the plant. As the cheese is made and moves down the line, everyone does a good job in perfecting the product all the way around."

That's true because the Gruyere received a judges' score of 99.50 — as close to a perfect score as a cheese can get. But Workman, too, was part of the crew, and he accepted the Grandmaster Cheesemaker title of the Governor's Sweepstakes Competition for his 30 employees.

Workman is a Monticello resident and dairy executive of Roth Kase USA, on Highway 69 on the northwest edge of Monroe.

This 41-year-old grandmaster now carries on the family tradition started by two Roth brothers in Europe and later expanded with a move to the New York and New Jersey area where they were joined by an associate — a cousin — who had been in the cheese business for years.

Their Monroe plant is a special business because it is designed so the public can take free tours. The plant and the 30 men and women who work at Roth Kase make 16 varieties of cheese.

Just what does it take to become a champion cheesemaker?

Bruce Workman

Workman, the son of a minister who grew up in the Monticello area, learned the cheese business as a young employee of Northside Cooperative, where he started working in 1971. He worked in the plant early, went to school, and returned to work later in the afternoon. He learned the business of cheesemaking, since the plant produced several varieties, such as Swiss, Muenster and Brick.

As a fast learner, he received his cheesemaker license in 1973 and worked for two cheesemakers. He advanced further when he bought a share of the business and later sold it to fellow cheesemaker, Paul Rufner, thus completing 17 years at Northside Cooperative.

That sale enabled him to join Avenmore in Monroe and later, the firm's Kent, Ill., plant. He added to his cheesemaking knowledge and skills with such varieties as mozzarella, provolone, Romano and a special Spanish variety — a white cheese.

Everything changed last December, when Roth Kase was looking for a Monroe plant manager. Workman applied and was hired for the top job.

State Crowns Young Cheese Champ

August 7, 1994

WEST ALLIS — At 33, Ron Sullivan, the youngest cheesemaker at Platteville's Old Wisconsin Cheese plant, is "definitely Irish" and now recognized as tops in the Wisconsin cheese industry.

In a ceremony Thursday night at State Fair Park, Ron was accorded an exceptional honor for a comparatively young cheesemaker.

Ed Eberle, who has run the sweepstakes contest for 36 years, said Ron is probably the youngest cheesemaker to win the top sweepstakes honor.

Ron makes Swiss cheese for Old Wisconsin Cheese, a family owned firm.

His cheese entry — a unique 10- pound wheel of Swiss — was judged at a near perfect score of 99.125. Ron now is Wisconsin's grand master cheesemaker and top winner of the 1994 Governor's Sweepstakes Cheese Contest.

And he wears his cheese crown well. He is soft-spoken and recognizes his superiors for giving him a chance to express his talents at the cheese vat and thanking the cheese plant's 99 dairy farmers whose high-quality, high-protein milk is produce on farms in Grant, Lafayette, and Iowa counties.

As he accepted his trophy, you could feel Ron was not alone as a champion but was accepting the honor for the entire team at the plant and on patron farms.

Ron Sullivan

And what was Ron's cheese worth when the wheel was sold in the cheese and butter auction?

Pabst Blue Ribbon representative, Bob Milkovich, bid $130 a pound, which means the 10-pound wheel brought $1,300.

Ron, a native of the Cuba City area, where his parents had a small plot of ground, has been making cheese at Old Wisconsin for a lucky 13 years. He got his start in cheesemaking when he was 18. He worked at the Carr Cheese Factory under Eldore Hanni after finishing Cuba City High School.

Ron and his wife, Sue, are raising three children in Cuba City and his precious time is spent with his family. He is a hunter and is taking a "tentative" turn at golf.

The auction raised $11,874 to cover the Sweepstakes Expenses.

The winners of sweepstakes class honors were Jonn Kovach of Land O'Lakes, Kiel, Cheddar, $2,300; Randy Marten of Cloverleaf Cheese, Stanley, Monterey Jack, $520; Milfred Severson of Klondike Cheese, Argyle, Brick, $550; Dean Polzin of Stella Foods, Almena, Italiana Grana, $900; Earl Wilson of the Burnett Dairy Cooperative, Grantsburg, Mozzarella string, $1,000; Raymond Kracht of S&R Cheese, Plymouth, Provolone, $750; Henry Zibrowski of Mindoro Creamery, West Salem, Bleu cheese, $1,200; Greg May of Kaukauna Cheese, Appleton, Snack Cold Pack, $280; Jeffrey Laack of Laack

Bros. Creamery, Greenleaf, Garden Vegetable, $125; Jamie Fahrney of Chalet Cheese Co., Monroe, Baby Swiss, $900; Ricky Rufer of Roth Kase, Monroe, Open Class Gruyere, $700; and Craig Straseske of Bresse Bleu, Watertown, Goat Milk Cheese, $262.50.

Butter sweepstakes winners were Louis Hebert and Brian Gwiazda, both of Grassland Dairy, Greenwood, $600 for salted butter and $450 for unsalted butter.

Dairyman Aims to Profit from Direct Cheese Sales

February 9, 1992

Gary Muench is a Lancaster-area dairy farmer — with three years' experience in direct farmer-to-consumer cheese marketing — who never gives up when he thinks he has a good idea going.

In fact, he has tested retail cheese sales in large and small business outlets in Illinois, Alabama, Georgia, North Carolina, Kentucky, Tennessee and Florida.

All this started in 1987 when Muench and scores of other Wisconsin farmers donated 200,000 bales of hay to drought-stricken farmers in the South, particularly Alabama.

This led to cheese sales in 1988 as Muench was encouraged to meet the demand for Wisconsin cheese in Alabama and in a network of farmers markets in other states.

Muench created the firm called Haymakers Specialties. With support of Southern farmers, food firms and government officials, Muench bought a refrigerated truck and went to work serving an expanding market.

Now, he said, it is time to expand his one-man show with support of Wisconsin farmers. He believes farmers could benefit from a new source of income at a time when every penny counts.

Muench, who is getting close to 50 but seemingly has the energy of a man in his early 30s, made his pitch for dairy farmers' financial backing as well as business cooperation recently at the regional meeting of the Wisconsin Milk Marketing Board in Madison.

He made his appeal as the last person on the program and didn't start until well past 3 p.m. — when dairymen are itching to head home to start milking.

This time, however, Muench's ideas kept most of the crowd of 400 farmers and their wives in their seats. He kept the message short while helpers distributed an information sheet on his marketing plans. Muench speaks and writes in plain language, warning that farmers are a "drowning breed" — being drowned by "government bureaucracy, red tape and big industry."

He said he has the organization to accomplish marketing goals with incorporation of Wisconsin Farmers Market Inc., which was formed with the help of an experienced farmer-supporter, Madison attorney Aubrey Fowler.

Muench describes the new organization as the first independent farmers group to produce the milk for the cheese and to promote and price its products before they are put in stores.

He said it is time to save the family farm by having farmers put $1 each behind the move so there will be the resources to stop "this drowning" through increased cheese sales.

The money will go into one fund to show that farmers are serious salespeople, and it should be eligible for a matching grant from the state.

"The (cheese) market is there if we apply our capabilities as independent farmers in the farmers markets, which will return dollars and "sense" back to the farm.

When farmers asked if the sales idea works, Muench telephoned an executive with Winn-Dixie, the largest supermarket chain in the South, to get figures on sales for Cheddar cheese carrying a bold orange circle with the label "Wisconsin Super Block."

Since the special Wisconsin label was put on the firm's packaged Wisconsin cheese, according to the executive, sales have climbed 5 percent. Muench said it is a good start, especially since Cheddar cheese sales were down, compared with other cheeses.

Now, with help of the Wisconsin Milk Marketing Board and the state Department of Agriculture, Trade and Consumer Protection, Muench believes that consumption increases will come to other dairy products marketed with that "close-to-home feeling."

Advocate Urges Communication

November 18, 1990

SUN PRAIRIE — Deborah Van Dyk could help young people who are trying to pick a career.

When she was active in 4-H clubwork, living on a dairy farm in Brown County and a student in DePere High School, she had a goal of being a communicator tied into the legal field.

During the observance of Wisconsin Farm Women Week at the Colonial Club Senior Center here, she told how she used 4-H and her education to work for what she wanted in her professional life.

She studied at UW-Stevens Point and UW-Madison. She is a writer, speaker and for those special expressions, a poet.

Today, she is director of Legal Affairs for Schreiber Foods in Green Bay, which is the largest private label cheese manufacturer in the country.

She is not a lawyer. Instead she manages the legal affairs, government relations and regulatory information work for a major company.

In addition to its Green Bay cheese shredding, cutting and slicing operations, Schreiber also has bacon and cheese processing plants in five other states. Even with its large sales volume, people seldom have heard of Schreiber Foods because it packages food as the house brand for its customers and only rarely under the Schreiber brand name.

She likes this particular business situation because Schreiber is closely allied with agriculture, the general business community and consumers. She also works closely with government agencies that set industry regulations.

Van Dyk is a well-organized woman who isn't a bit reluctant to step into two of today's biggest policy areas affecting farmers, consumers, business and government — food safety and solid waste recycling programs.

All this has happened because she fully appreciates what is going on with the consumer, farm and business sectors of the economy.

As a single parent, she is raising two children, a 12-year-old daughter and a 9-year-old son.

To keep up with the busy world of technology, government regulations and changes in consumer perception, Van Dyk does a lot of reading. A lot of it is done at home with her children.

"When the kids are doing their homework, I do mine and read. I think it is a good work ethic," she added. She describes life as a single parent as "wild at times, but I love my kids and wouldn't trade them for anything. When I have to be away my family is there with the children."

At a time when some consumers have trouble trusting government programs like the Food and Drug Administration in food safety and even challenge the efficiency of solid waste management and recycling, Van Dyk wades right into the problem-solving phase of each program.

She is a member of the Wisconsin Legislative Council

Deborah Van Dyck

committee for Solid Waste Management, is an interpretive and informative speaker on food safety and chairperson of the legislative committee of the Wisconsin Dairy Products Association.

What she wants to see is straight-forward communication between consumers and the decison makers in agriculture, business, and government who are charged with carrying out legislative and adminstrative policy.

"If we communicate with people they have a chance to understand the issues," she said.

"We are in an age of communication, which is such a key element to everyone. If people understand the issues they can make an informed choice. If they don't understand, there is absolutely no chance for people to make an informed decision."

In food safety, Van Dyk's example of consumer confusion involves the advice of health care professionals to reduce fat consumption and increase the intake of fruits and vegetables to enjoy a healthier, longer life.

At the same time, consumers have been told by some groups their apples are coated with Alar and high pesticide residues are found on fruits and vegetables, that salmonella contaminates tomatoes and chickens, milk has antibiotic residues, fish carries PCB's and "a staple in my son's diet — peanut butter — has aflatoxin.."

"It's no wonder consumers are confused about the safety of food they eat," Van Dyk added.

As for Alar, now that emotions have died down, the

rational scientific truth about the compound is being heard, she said. The truth is that the documents used to cause Alar hysteria were based on a study discredited by the National Cancer Institute and by a federal scientific advisory panel review, Van Dyk noted.

"There is no scientific reason to ban the use of Alar," she said.

Having the safest food supply in the world, is not a fluke. Food safety is based on scientific review and practice, Van Dyk continued.

Technology is readily available and more and more of it is coming.

"In the past three years I've seen the fastest change in industry since I started 11 years ago. I don't think it is going to slow down, especially since consumers have become more and more vocal," she said.

Chinese Visitor a Farm Booster

November 25, 1990

The new census in China shows the population at 1.1 billion and a few of that number are in Madison as students and scholars.

But a special one in the delegation is Jen How Wang.

Jen How Wang has taken a strong interest in Wisconsin agriculture. Wang, 69, who could easily pass as a man in his early 50s, is a Chinese professor of electrical engineering. He also has a broad knowledge of Chinese agriculture and says China needs American agricultural technology and equipment because it is technologically backward. Of its 1.1 billion people, China has 800 million persons in farming, he said.

In 1987 and 1988 he was a visiting scholar in the UW-Madison College of Engineering, where he specialized in superconductivity. He returned to his home in Shanghai when his teaching contract was completed.

But he wasn't home for long.

Wang is back in Madison visiting his daughter, Zian Wang, who is a doctoral candidate completing her degree in electrical engineering.

The academic field of electrical engineering runs like a current in the family. Wang's wife, 72, also is a professor of E.E. as is his son-in-law Ju Wang, who has his doctorate from the UW and now is working at the Argonne Laboratory in the Chicago area.

Wang said with a laugh that he is now a baby sitter for Alice, his 4-year-old grandchild.

"But now I am free. The time is all mine, now that I have been retired from teaching for two years," he said. Each morning he jogs two miles, which he said keeps him "in shape and young."

Wang has an excellent knowledge of English, both speaking and writing.

He learned English from American troops as a member of the signal corps in a liaison group of Chinese Army personnel. He serve in Burma under General Stillwell from 1944 to 1946.

He returned to China in 1946, was discharged and for a time worked in a customs office in Shanghai and later in an aircraft factory in Nanking.

In 1947 he was married and he and his wife have two daughters and a son. He was educated at Southwest University in China and graduated in 1944. Most of his professional life was spent as a professor of electrical engineering in a Chinese university.

Now that his time is his own he has had a chance to think of how close China and the U.S. have become and what must be done to improve that relationship.

Wang is an advisor to a science and technology commission in Zian City to work out a strategy for the development of agricultural, industrial and education programs.

He is particularly sympathetic to the price drops Wisconsin farmers and cheesemakers have been going through and immediately thought of how quickly the increased cheese production could be used in China.

"Yes, I like cheese and most people in China like it, although tastes are different," he said.

Incidentally, his favorite food in America is pizza and Wang says he likes to take his cheese as part of the pizza, which is a once-a-week treat for the Wang family.

He learned to eat and appreciate cheese — a processed American cheese — because it was part of the American Army rations in Burma.

But he does not think in terms of importing Wisconsin cheese for Chinese consumption. The China policy, instead, is to plan joint operations with cheesemakers.

It will be a long term process, however, since China will have to have greater cow numbers before it has the source of milk for cheese production.

What a joint operation in China means is that China provides the land, buildings and labor for cheese operations and American investors provide the equipment and technology for production, packaging and distribution.

Wang said that China now is coming back to normal as a place for economic investment.

"Most foreign countries are coming back to China and only America is being left behind," he continued.

Jen How Wang

"As known to everyone, China is one of the developing countries. It has a great area to be developed and that should attract the interest of the global community," Wang said.

He suggests that people of the world should have the vitality to exploit the virgin lands of China under a government arrangement as the U.S. did in the American West a century ago.

Wang noted the warm welcome China has given to people to develop its natural resources as well as projects for industry and education.

"Up to now most Western Countries have their interests in the industrial field, but no one is concerned about agriculture. In reality, agriculture in China is a weak, but promising area, which needs prompt renovation and mechanization," he said.

That's why China is looking to the U.S., since it has advanced agicultural technology and good managment.

He said this American investment in China with a technology transfer will become profitable and profits can be sent back to the U.S. as well as used in China.

"That is why I stubbornly suggest that agricultural technology, a strong item in Wisconsin, be transferred to China," Wang said.

Should anyone be interested in this," he said, "I'll gladly help them plan to cultivate the virgin lands in China."

State Fair Crowns Cheese and Butter Making Champs

August 3, 1996

WEST ALLIS — The brilliant August weather was the perfect setting Thursday for the annual Governor's Sweepstakes Cheese and Butter Auction at the Wisconsin State Fair.

Gov. Tommy Thompson's enthusiasm helped push the

total sales figure on the 14 lots of cheese and three lots of butter to $12,367.90.

The money will be used to educate fairgoers about Wisconsin's dairy industry, according to Timon Zander, president of the State Fair Dairy Promotion Board.

Dave Henning, cheese grader at the Land O'Lakes Kiel plant, turned cheesemaker for this contest and produced a 42-pound Cheddar cheese. Judges gave it a remarkably high score of 99.625 points — highest in the contest — so Henning was honored as the 1996 Grand Master Cheese Maker in the championship showdown.

James Hanson, president of Volm Bag Co., Antigo, a major supplier for the cheese industry, was a bit flustered, but he stepped up to set the day's auction cheese price at $75 a pound.

For Land O'Lakes, it marked the 15th straight year their Cheddar cheeses topped the class of those popular cheeses. And the $75-a-pound mark was continued for three other cheese entries, earning $3,150 for the promotion effort.

Peter Zander of Zander's Creamery, Cross Plains, took two of the honors in butter competition, capturing the trophy crowns for their entries of unsalted butter and flavored butter.

Bids of $60 a pound gave the firm credit for earning $1,200 for butter promotion at the State Fair. Last year a trio of Zanders swept all butter classes.

The unsalted butter of the Governor's Sweepstakes gained a high bid of $60 per pound, made by Dairychem of Noblesville, Ind. Dairychem makes butter ingredients used by Zander's. Jerry Dryer, dairy industry marketing leader and analyst, had the $60 a pound bid on Zander's garlic-and-herb flavored butter entry.

Zander's entries were scored at 99.25 and 98.6 and brought $600 each.

Eric Gunderson, 25, entered his 42.5-pound goat milk Cheddar cheese and won the class with a score of 98. He

Gov. Tommy Thompson and Dave Henning

is cheesemaker for Mount Sterling Cheese Cooperative, which makes goat milk cheese from producers in Wisconsin, Iowa and Illinois.

Gunderson was born in Boscobel, the son of Crawford County dairy farmers Phillip and Rose Gunderson. After he graduated from North Crawford High School in Gays Mills, he decided to take a job at the Mount Sterling Cheese plant.

Gunderson's goat milk product was purchased for $7 a pound, adding $297.50 to the promotion pool.

Farm Background Helps Insure Quality

March 18, 1990

For a farm boy from an Ixonia area dairy farm to be manager of the only traditional cheese plant in Madison is just about the best thing that could happen to Leroy J. Schumacher.

Schumacher, 51, has held the job for 10 years and runs the cheddar cheese and whey concentrating operation at the Morning Glory Farms plant of Associated Milk Producers Inc.

Each day the plant turns out more than 30 tons of cheddar cheese and two large tanker loads of concentrated food grade whey.

Schumacher worked on the home farm until he finished Ixonia high school and then, at 18, started working in dairy manufacturing plants, including the Pabst Farms plant in Oconomowoc.

He became a skilled manager of a variety of dairy operations through on-the-job training under an expert, like Earl Janssen, now a dairy products broker who for years was with Pabst Farms at Paoli and Oconomowoc, and later, AMPI in Madison.

"I guess my farm background really helped me appreciate work and how important quality production is in marketing. I was able to make the most of it when I got a chance to join Pabst Farms," Schumacher said.

"But my real interest was in cheese production, something I enjoy. It is definitely a challenge, there are new technologies coming along all the time and it gives a person an opportunity to use these technologies to achieve high quality production," he added.

"I got to know many people in dairying and by listening to them I got my dairy education. This allowed me to work my way up through the ranks," he said.

AMPI is a dairy cooperative and the Madison Morning Glory Farms plant is part of a regional AMPI operation, which serves as the milk processing and marketing arm for 1,300 dairy farmers in 25 counties of southern and west central Wisconsin.

A large portion of the daily milk supply from those farms, about 675,000 pounds of Grade A milk, is moved into the Madison operation.

But most urban people probably know the Park Steet plant more for its retail cheese store with 40 cheese varieties and its year-round gift cheese operation.

Now the annual operating schedule of the plant is changing, Schumacher said.

It used to be operated as a "balancing plant," which is a dairy industry term to describe a plant that produced dairy products about six months of the year and then closes from July or August until January.

That's a period when milk production is low and it is more profitable to use Grade A milk as bottled milk than to use it for cheese manufacture. At this time the milk is diverted to markets in the South and Southeast U.S.

However, the Madison plant now has a new contract with Schreiber Foods Inc., Green Bay. Schreiber will buy all the daily cheese production and move it to its processing plants in Green Bay, Wisconsin Rapids and Carthage, Mo.

This attractive contract, as Schumacher terms it, is based on quality cheese production that is available in large volume.

That means the plant will operate on a 24-hour, seven day a week production schedule, making cheese for the entire year. For 20 hours the plant will be producing cheese and the remaining four hours will be used to clean the equipment. There are 24 persons employed in the plant cheesemaking and whey condensing operations.

The cheese produced at the plant is called barrel cheese, a cheddar sold in curd form in fiber barrels. Each barrel holds 500 pounds of cheese.

Barrel cheese is used for making process cheese, which has a high demand from restaurants where it is used in cheeseburgers. It is also used for cheese spreads.

Whey is a by-product of cheese manufacuture. Each 10 pounds of milk yields a pound of cheese and 10 pounds of whey.

At the Madison plant the whey is condensed by taking off the water so the whey can be moved in two highway tank trucks rather than in the 12 tankers it would take to haul the whey in natural form. Condensed whey contains 42 percent solids in contrast to the 6 percent solids in natural whey.

The whey is moved to Foremost Foods for use in a wide variety of food products.

Cheese Champion Keeps Victory in Perspective

August 8, 1993

WEST ALLIS — It was time for the Governor's Cheese Sweepstakes at last week's opening of the Wisconsin State Fair.

Ivan Gobeli, 57, of Franklin Cheese Cooperative factory, Monroe, was named the 1993 Grand Master Cheese Maker for his excellent Muenster Cheese scored at 98.63.

What kind of a price will a high scoring Gobeli Muenster bring?

The crowd of bidders at the sweepstakes auction bid long and hard for it until Pabst Brewing won with a high bid of $135 a pound, pushing the total price to $2,632.50.

Gaining the top spot in the showdown competition with 13 other class winners in Wisconsin was a triumph for small cheese cooperative factories like Franklin, which appear to be vanishing from the Wisconsin countryside.

But when Gobeli became the grand master this year it also marked a double win for the now famous Franklin Cheese Cooperative factory, located west of Monroe on Franklin Road, a mile south of Highway 11.

Last year, Gobeli's son, Steven, set the sweepstakes pace for the factory when he won grand master cheese maker honor, also with a Muenster.

"It's nice to win," the elder Gobeli said, "just as long as you don't let it go to your head. You can keep that from happening by keeping in mind that tomorrow morning you have to get up and make cheese.

"And remember we won with cheese made from the high quality milk from the 16 farms in the cooperative that ship 40,000 pounds of milk a day to the plant."

Gobeli had a short celebration as grand master at the fair because the next morning he had to haul milk from the farms to the factory.

Gobeli has a good reputation with his milk producers because he knows the farmers spend long days getting their work done and they also expect him to spend some long days at the factory.

He has cut back some with his son Steve and his wife, Janet, at the factory each day. Cheese making means long hours — like getting up at 2 a.m. each day to turn and wash 200-pound wheels of Swiss cheese. He has many 10 to 12 hour work days to his credit.

He relaxes, though, by breeding and raising exotic birds, such as parrots, small and large, as bird lovers order them for their own enjoyment.

"And when I married him (in 1955) he had a parakeet," Marlene said.

The Gobelis, both Swiss, met in nearby Franklin School. The couple was married before Gobeli spent two years in service and before learning the cheese business. They have three sons and a daughter.

Gobeli and Marlene are partners in the factory management, where Gobeli started as an apprentice in 1957. Partner Marlene handles the factory books.

He explained that in a cooperative, the farmers own the building and the Gobelis own the cheesemaking "tools," hire the labor and market the cheese.

The Gobelis started acquiring the management part of

Marlene and Ivan Gobeli

the operation in 1962, when one of two partners sold his share in the business. A few years later the other partner sold out to the Gobelis.

The unique part of a cooperative operation is that the monthly milk checks to the 16 dairy farmer owners is determined by the total income the cheese commands on the market.

Occasionally, one of the milk cooperative owners will leave, but Gobeli said, there always has been another dairy farmer to take the place of the departing producer.

He is proud that a few of his producer families are the third generation shipping milk to the plant.

There are eight such cooperative plants in Green County, the highest number for any county in Wisconsin, and two cooperative cheese factories in Lafayette County.

The Franklin Cheese factory traces back to 1890.

New Kind of Dairy Farming May Ease Price Pressures

September 6, 1992

There is a new kind of dairy farmer emerging in Wisconsin's dairy country.

Jerry Heimerl, of Cleveland in Manitowoc County, who runs a 150-cow dairy herd, is in the final phase of deciding whether he should add a farmstead cheesemaking operation on the home farm.

He is trained as a cheesemaker by the Food Science Department and the Cheese Research Institute at the UW-Madison College of Agriculture and Life Science.

Then he chose Dan Carter, Mayville, national and international expert in cheese marketing as a consultant to get his marketing inspiration and skills.

And the State of Wisconsin, through a special grant program administered by the State Department of Agriculture, Trade and Consumer Protection, provided Heimerl with the financial aid needed to become a full dairy entrepreneur — both as a milk producer and a cheesemaker.

Heimerl was the first dairy farmer to use the grant program and he was given two grants — one for $24,000 and the other for $10,000.

He has his own cheese specialty, a Port Salut style.

He chose this route to farmstead cheesemaking because he was worried about the constant pressures by milk markets and the government to push milk prices below the cost of production.

He said the first decision he faced was to get bigger by adding more cows, buying more land and purchasing more equipment and therefore, taking on more debt.

But Heimerl rejected that because his big concern was how it would change his family lifestyle.

At the same time he was thinking of what happens to the 10,000 pounds of milk he sends to market each day, since it goes out the driveway without him knowing what happens to it.

"Actually, we forget about it, turn around and start milking the cows again."

One thought returned to him over and over. There has to be a way to put a value on that milk to counteract the pressure to push down prices paid to dairy farmers.

He wonders where milk prices are going — "even though we feel good about it right now — but come December, who knows?"

"There has to be something to turn this around and I am looking at sales or a different way of marketing. I figure we have endless opportunities if we use our own milk and that brought me to the decision of what I can do with my own milk — make it into cheese," he said.

He knows the risks and only if a feasibility study meets the standards of his banker can he expect to receive the financing of $250,000 to $300,000 for his farmstead cheese operation.

"When we think of the risks, we sometimes think we are standing in a quagmire," he said.

But he has confidence, he said, because his consultant, Dan Carter, has made the market analysis showing that there is a place in today's market for specialty cheeses of high quality.

Cheese of high quality has strong consumer demand and gains a premium price.

The estimate from the American Cheese Society, which represents specialty cheese producers, is that Wisconsin could have 20 or 25 such operations in the next 10 years.

Just how big is an investment of $300,000 or more?

In terms Heimerl knows best, "It is what you would need to buy a 150-cow Holstein herd with a rolling herd average of 23,000 pounds per cow," he said.

Heimerl knows dairying. He was born and reared on a dairy farm in the Brownsville area of Dodge County. He enlisted in the Navy for four years during the Vietnam War.

When he was discharged, he used the GI Bill to enroll at UW-Stevens Point and graduated in natural resource management.

During that time he met Elise Klessig of Cleveland, who also was studying at Stevens Point.

After graduation, they were married in 1978 and a year later, Edward Klessig, his father-in-law, asked Jerry and Elise if they would join in the Klessig family operation. They now have a partnership arrangement for management of the Saxon Homestead Farm.

Edward Klessig is 73 and is at the farm every day.

Klessig has a close relationship to Madison because he was the farmer who brought a Holstein herd to Madison and kept them, milked them and cleaned them daily on the State Capitol lawn to protest the routing of I-43 through his farm.

Beaver Dam Business Longtime Attraction for Cheese Fans

July 2, 1995

BEAVER DAM — On Saturday there was cause for celebration at the Schultz Cheese Haus, as Harvey and Nancy Schultz started their 18th year in the cheese sales business.

Their Cheese Haus, just south of Beaver Dam on Highway 151, is really a service business, a Wisconsin institution in which there are strong bonds between customers and the Schultz entrepreneurs.

There even are customers who ask Harvey and Nancy not to talk of retirement because they depend on their help and recommendations, whether they are buying Wisconsin cheese or selected Wisconsin wines.

But Harvey and Nancy are considering, even anticipating retirement. It all depends on the sale of their business.

"Our business is doing very well, but we are getting a little tired," Harvey said, noting their business is a seven-day-a-week operation.

Harvey is a native of the Beaver Dam area and he once owned the City Dairy, which served customers in and around Beaver Dam. He sold that and then trained in accounting and business law.

Nancy's home was in Long Beach, Calif., where the couple met and married when Harvey decided to settle in California a few years after he completed his four years with the Army.

He served in Army finance offices in Germany, during the Berlin Airlift in the late 1940s and later in the United States at the start of the Korean War.

In California, he was with a wholesale cheese business and other firms, which depended on his accounting and management skills.

After making a family visit to Beaver Dam, the Schultzes returned to California, but for three months talked about moving to Wisconsin.

Finally, it was Nancy who encouraged the move to Wisconsin "to get away from the California rat race."

They came back to live here because a cheese sales business was for sale and

Harvey and Nancy Schultz

that was what they wanted for their future in Beaver Dam.

Sale details were being completed, but the owner suddenly changed his mind and decided not to sell. Fortunately a friend had learned of another cheese business seeking a buyer and that's when the Schultzes acquired their current business home.

In the intervening 17 years, the Schultzes sold "tons and tons" of 87 varieties of cheese featured at their prized Beaver Dam Highway 151 location.

Not only did it cater to people of the Beaver Dam area and other state customers, but their Cheese Haus has been the annual stop of summer vacationers on their way home to Illinois and other states.

The business also serves the crowds returning on Highway 151 from the Experimental Aircraft Association's annual fly-in at Oshkosh. And for some bus lines it is a regular stop for their tour groups visiting Wisconsin.

The friendships with customers continue for years and some of the elderly will suggest to Harvey and Nancy they are getting too old and won't be able to come and get their cheese anymore.

But, even though the Schultzes don't know the names of all their customers, they recognize their faces as their customers come in and happily announced, "well, we are back again."

It makes you wonder which of the 87 varieties of cheese the elderly are eating to enable them to come back year after year.

When Harvey is asked which is his favorite cheese, he pauses for a moment and says, "medium aged brick" — his choice which was originally developed in Dodge County.

For Nancy, it is sharp cheddar cheese as a favorite.

Fortunately, she has plenty of choices because Harvey includes cheddar cheeses that are 5 years old with the degree of sharpness she enjoys.

The big cheese time of the year is getting the hundreds of gift cheese packages packed and sent throughout the United States for Christmas.

Harvey said cheese sales are showing a general increase, which he attributes to the promotional cheese programs of the Wisconsin Milk Marketing Board, an agency financed by money from every Wisconsin dairy farmer.

He has kept a close watch on WMMB activities and what they contribute to the sales of cheese and dairy products, since he served a term as president of the Wisconsin Cheese and Specialty Food Merchants Association of Madison.

Chapter 4

Dairy Business

Dairy Expert Valued Source

November 15, 1992

Reedsburg native Jerry Dryer has to be extremely careful should he ever decide to trim back his business, known as the Jerry Dryer Group.

Such cutbacks in business are popular today, but in Dryer's operation, he could be the first one to go because he and a full-time secretary are the only employees.

Dryer, the owner, president and chief executive of the Jerry Dryer Group in Northbrook, Ill., is really a one-man show. As a consultant, Dryer helps dairy plant operators and marketers make product and operational shifts to improve profitability.

When he gets a call from a client with a problem, Dryer will find a specialized consultant or two to form a team to work with plant executives to find a profitable course of action.

Dryer now is making waves in the rapidly changing dairy business, by working on marketing, product development and issues management to keep track of trends and how to prepare for them at the plant level and on through to the consumer.

He turns out volumes of words in articles and features and publishes a price-forecasting letter for a limited number of clients. He also has a weekly radio program going to 80 stations.

He has clients from California to New York and has become a frequent flier so he can see what's happening on the spot.

Dryer, who is divorced, takes time off to visit his three sons for whom he has set up trusts for their college educations.

He got enough money for the trusts by figuring out a way to create effective dairy-information competition at the right time.

While in Madison he started two publications, one on cheese and one a dairy letter. When his words and growing circulation started bothering a larger Chicago-based dairy publication of Gorman Publishing, they came to Madison to buy out Dryer and made him editor of their principal dairy foods publication.

Dryer was raised on his father's Sauk County farm off Highway 23 between Reedsburg and Wisconsin Dells.

The farm had a small herd, which gave Dryer a dairy introduction. He got an idea how qualified consultants worked when Bob Greffert, a Reedsburg Hereford breeder, helped the Dryer family set up a purebred Hereford herd in the 1950s.

Dryer graduated from Webb High School and studied agricultural journalism at UW-Madison, where he received his master's degree after working as a research assistant with agricultural economists Frank Groves and Richard Vilstrup.

Jerry Dryer

After completing his studies in Madison, he went to Washington to join the National Milk Producers Federation, which he called a logical starting point.

He said there is enough going on in Washington that affects the dairy business "and it serves you well."

Dryer also was with The Cheese Reporter in

Madison for a year, to work with publisher/editor Harry Palmiter and get acquainted with Richard Groves, who is now editor and publisher of the publication.

He reflects on how much he has seen of dairying, going back to the days when milk supports were increased every six months. Those supports led to the diversion programs to cull cow herds to reduce production and later the dairy termination program, which took entire farms out of milk production.

Dryer still believes dairy farmers could help themselves through legislation if they get to the point where they can sing from the same hymnal. Even the National Milk Producers Federation can help them legislatively.

"But dairy farmers have to face political reality since there are only about 150,000 of them at a time when there are 100 million voters," he said.

"I think there is a place farmers can have a voice — through their cooperatives and promotion groups," he added.

"You have to remember that when Congress started lowering the government support price on milk, it gave them the producer-financed promotion program so they could work to get their price return out of the marketplace and not the federal treasury," Dryer continued.

But the big opportunity for farmers and dairy plants to capitalize on their business is not too many years down the road, he said, when larger and larger dairy product exports will be made to the world — including Europe.

Farmer Takes Positive Approach to Marketing

April 14, 1991

In a real sense, Timm Johnson, New Richmond dairy farmer, is the only person in a dairy organization to represent all of Wisconsin's 33,000 dairy farmers.

He is president of the Wisconsin Milk Marketing Board, the Madison-based promotion, research and consumer education organization, to which every dairy farmer contributes (through his milk check) to the board's $25.7 million annual budget. The 1991-1992 proposal was set last week.

Johnson, 40, and his wife, the former Helen Thiede of Columbus and their two daughters, Courtney, 16, and Melissa, 14, have a 200 acre dairy farm with a 40-cow registered Holstein herd.

The Johnsons have been on that farm since 1972 when he graduated from the UW-Madison in dairy science. He has been active in his county and community, having served as president of the Polk County Holstein Association, a member of the Wisconsin Indianhead Technical Institute advisory board, including a term as president, and president of the Polk County 4-H Leaders Federation. Currently he is a member of the Osceola School Board and serves as its president.

He was elected by dairy farmers as a director of one of the state's 25 milk marketing board districts in 1987. The district directors on the board, each representing dairy farmers in a single or multi-county district, elected Johnson president in 1990.

The milk marketing board, created by the Legislature and elected by dairy farmers in 1983 has a state mandate "to expand and maintain domestic sales of milk and dairy

Timm Johnson

products, develop new products and new markets, improve methods and practices related to marketing or processing of milk and dairy products and inform or educate consumers."

Johnson said he believes the board is making an impact on economic welfare of the dairy farming.

Johnson got into the milk marketing system when a group of dairy farmers in Polk and surrounding counties asked him to run for the district directorship and represent them on the board.

The board presidency is a position he did not seek, but "it's just something that happened. It seems that I just end up with these positions. I feel comfortable as president.

"What motivated me is the sense of accomplishment a person can experience in being able to make an impact on the dairy industry," he said.

"With milk at $10 per 100 pounds there really are a lot of negative factors on farms," Johnson said. "On our farm in Polk County we are dairy farmers, so we are aware of this. At the same time, we see a lot of really positive things happening on a step-by-step basis."

Johnson doesn't see any one major accomplishment that solves the problem of $10 milk, the lowest price since 1978. Instead there are going to be a number of smaller things — the little things that will meld into one and "hopefully we can solve the problem."

He appreciates the opportunity to try to help steer the milk marketing board to be part of the solution and help make "the Wisconsin dairy industry stronger. That's my motivation."

The WMMB experience is an in-depth course on dairy marketing and Johnson said the first thing that strikes a dairy farmer elected to the board is how complex the entire dairy industry is.

In most cases the dairy farmer is a milk producer and once his milk leaves the farm he believes his job is done.

But after the farm there are many persons involved — handlers, processors and marketing people needed to transform farm milk into finished products on a consumer's table somewhere in the nation or virtually anywhere in the world.

Programs of the marketing board are changing with the rapid changes in dairying and one in particular Johnson sees with potential is a joint ventures program with a dairy plant that may be willing to try new systems and products with a partner like the WMMB.

The new budget has a $400,000 joint ventures item to expand new lines in the dairy industry. "We realize margins are really tight for plants on a lot of products, but if someone wants to start a new line, in product packaging, presentation or promotion, we, on the board, could share some of the risk with the dairy operator and do something together to advance a new trend in dairy products.

"We can get the notion of new dairy methods started and then its up to industry to take advantage of it and the profits it could generate," Johnson said.

"The opportunities are endless. We have to find them and then put the dairy industry in position to take advantage of the new products and systems," Johnson said.

Dairymen are becoming convinced they cannot live only on fluid milk, butter, cheese and nonfat dry milk. The expanding base of dairy income may be in milk components, used not only in food, but industrial products, Johnson explained.

"I see a lot of farmers and they are positive. You have to be positive because you will go nowhere if you are negative," he continued.

"I believe very strongly you must look for opportunities in every problem instead of looking for the problems in every opportunity. That's the basic philosophy I have," Johnson said.

But he also has great admiration for the support he is getting from his wife and family when he has to be away from the farm.

"My wife and daughters take care of the day-to-day stuff along with a student from the vocational school in New Richmond. This is very encouraging to me. They have just been superb," Johnson said.

Beck Left Mark as Artificial Insemination Tech

February 3, 1991

Don Beck is alone at the top of his profession. No other person in the world has gained the status that Beck achieved in more than 40 years as a technician skilled in the artificial insemination of dairy cows.

Beck's achievement stands at 217,770 cows bred, which his longtime employer, Tri-State Breeders Cooperative of Westby and Baraboo called an "unequaled career achievement."

He has a plate hanging on the dining room wall with a painting of the cow that was the 200,000th cow he bred, a daughter of the bull named Bubba.

In 1967, he reached his first milestone as a technician when he bred his 100,000th cow.

Now someone certainly is going to have to get the forms from the people of "Guiness Book of World Records" to get the name of Donald Henry Beck established not only as Tri-State's person of the year, but as the international person as well.

Not only is he a great technician of the art of cow breeding, but he has the ultimate in public relations skills that made him a trusted friend of dairy farmers in Dane and some surrounding counties.

"You have to smile when you arrive at the farm and most times are greeted by the farmer. And you have to talk with him and listen to his thoughts since your last visit," Beck said.

Don Beck

To go with that, a technician has to love to work long hours practically every day of the week and month. Beck's wife, Dorothy, said the hours mean missing supper most of the time.

Today's technicians get a break and are off three days a month.

Beck is 65 and retired, but he still does some herd analysis work for Tri-State in its three-state area and in Mexico. And now, except for those occasional trips to farms, he is home for supper and Dorothy is happy. The Becks live on Madison's West Side. They are parents of two grown children.

Beck was born in Waunakee, where his father had a restaurant, which closed as the Depression hit businesses.

So his father moved to a Sauk County dairy farm northwest of Spring Green, where young Beck was raised in the ways of agriculture to learn about milk cows, hogs and crops.

Those were the important days of Beck's life because they helped establish his goal, to work in a field of agriculture. His Spring Green high school agriculture instructor, Roy Melvin, was a part-time technician and helped Beck become interested in artificial insemination.

Beck was serving with an Army Engineer battalion in Japan after World War II when he decided he would be an

AI technician.

When he was discharged in 1948, he not only married Dorothy, who is from a farm in Lime Ridge, but he attended a short course at the University of Wisconsin to train technicians. Those were the days when the UW ran its own bull stud as artificial breeding was starting and Beck was hired as a technician to work in southern Dane County.

With much of UW's artificial breeding and business research completed, the bulls were sold to a new cooperative formed in Madison, Southern Wisconsin Breeders Cooperative, and Beck made the move too.

In 1961, Southern merged with Tri-State Breeders and Beck was with his third AI firm. For a time he was given regional and herd analysis assignments or his total number of cows could have been larger.

Farmers shifted to artificial breeding mainly because it meant they could get rid of the bull.

"That bull on a farm was a pain — and dangerous," he said.

As he moved to the 100,000 and 200,000 bred cow marks, Beck often was asked if he set a goal.

"Yes, my goal was to breed as many cows as possible," the practical Beck said.

He once hit a high in cow breeding when he bred 76 in one day.

To get to the farms of his customers, Beck drove 1,890,000 miles. He lost track of the number of cars he had to buy, but in the old days he would drive them about 60,000 miles before getting a new car. In more recent years, the cars held up for 140,000 to 176,000 miles.

"Some people say the cars years ago were the best. But now the cars have motors that just go and go." He had his experiences, too. A farmer tied two heifers to be bred along Highway 14 near Oregon when the highway was the main route to Chicago and weekend travel was heavy.

"I bred the heifers as the line of traffic slowed and those

Chicago-bound travelers probably are still wondering what I was doing to those heifers. I was a little embarrassed," he said.

One thing Beck got acquainted with was all the farm dogs. Most were friendly so one Christmas, he decided to give the dogs a big, milk bone dog bisquit as a present.

"But I found out that those dogs remember and they pestered me on every visit following to get another bone. After that I always had to carry a milk bone to the dogs." he said.

Beck often thinks of how his customers are doing now and what dairying will be like in the future.

When he started as a technician in Dane County, milking herds averaged between 35 and 40 cows, but now they average 60 to 70 cows because farmers are better managers.

Beck believes that within five years all milk will be sold without butterfat being priced. In the future, pounds of milk and pounds of protein will be more important than pounds of butterfat, he said.

Woman Heads Dairy-Deli Group

September 8, 1991

Andrea Neu, (pronounced "Nye") is the vice president of a $24-million-a-year agency funded by 33,000 Wisconsin dairy farmers.

Her position is with the Wisconsin Milk Marketing Board, where she is reponsible for such key dairy farmer-financed programs as dairy product advertising, promotion and communications.

But she now is adding a voluntary committment to expand her partnership with dairy farmers and is serving as the first woman president of the Madison-based International

Dairy-Deli Association, a trade organization.

The IDDA has 750 corporate members from major dairy and food, manufacturing, processing and distributing firms along with individual buyers and sellers with dairies, delis, bakers and supermarket food services. Members represent firms throughout the U.S. and Europe, Scandinavia, and Australia.

Andrea Neu

Like the Milk Marketing Board, the IDDA is a "result-oriented" service organization and both are creating sales results for Wisconsin products.

The association originally was created in the Wisconsin Department of Agriculture by Willard Reese to move Wisconsin cheese into national markets.

But when the association expanded to other foods, Wisconsin appeared to be losing its dominant place as a cheese supplier and Neu and Jim Tillison, former executive director of Wisconsin Cheesemakers, talked to the IDDA board and were authorized to allow the Milk Marketing Board to set up a Wisconsin section of cheese manufactuers. This allowed the state to compete with overseas cheese firms, which had quickly moved in to serve U.S. deli markets.

Neu, from Green Bay, traces her dairy roots to Shawano County on the farm of her grandparents, where she spent her summers. This dairy experience led to her understanding and dedication to dairy farmers, which she carried to the American Dairy Association of Wisconsin as director of advertising and promotion, and the Wisconsin Milk Marketing Board. She was the second person hired by the board when it was established early in 1984 after farmer approval of the state's milk marketing order.

Neu was a veteran businesswoman when she joined the dairy promotion units. At first she wrote ad copy for her husband, Frank, a graphic designer and commercial artist in Green Bay, and then in Milwaukee.

When the couple moved to Madison, she recalled, "I realized it was my opportunity — my time — to go to school and earn a degree." She was graduated in 1974 with a degree reflecting her studies in communications, foods and nutrition, business and consumer science.

Her university education was possible, she said, because both her children, Patrick and Debra, were in the higher grades of Madison schools.

As an older student, she knew exactly what she wanted to be and the courses she had to take to achieve that goal. She wishes more people could have that chance.

She tells of students who were not sure what profession they wanted and were not certain as to what courses to take. It is something that doesn't happen to an older student, she said.

After her three years with the state American Dairy Association, Neu left and formed her own business, Consumer Medium, and worked with dairy processors with new products and provided creative services and marketing/communications programs for other clients.

Later she was consumer adviser for the Copps Corp. and learned about the retail and wholesale business. Neu believes there has to be a partnership between dairy farmers and organizations they create or support with their monthly assessments. The key is to work with business associations strongly based in dairy product marketing, she added.

In today's business world, she added, dairy farmers cannot be successful by acting on their own. The WMMB is the catalyst to make things happen in the dairy marketplace.

Partnerships established in the marketing channel, Neu continued, firm up relationships with brokers, distributors, retailers and food services right on to the end users — the market operators or consumers.

Now, what is done by the Wisconsin Milk Marketing Board in domestic markets to create additional dairy sales, is enhanced with the added market contacts through the International trade group Neu heads.

She has been connected with IDDA through the Milk Marketing Board since 1985, serving on various committees and the executive committee in preparation for her year as president.

IDDA Exec Discusses 'The Highway'

June 26, 1994

Madison's Carol Christison put more than 4,100 executives, managers, and operators on the fast track a few days ago in Baltimore.

Christison is executive director of the International Dairy Deli Association (IDDA), which is the focal point of new food business ideas.

This annual event started in 1963 as the Wisconsin Cheese Seminar, a creation of Willard T. Reese, Madison, who is now retired from the Wisconsin Department of Agriculture markets division. The seminar quickly grew to become the National Cheese Seminar, and as membership and interest zoomed, the seminar expanded to reflect the international scene, new members and global interest to become the International Dairy Deli Association. Christison has been the top executive for 11 years.

In her report to the food industry — ranging from suppliers to supermarkets and specialized shops — Christison said her fast track emphasis is the information highway that has the potential of changing the lifestyles, consumers and consequently the operations of food business to better serve their millions of customers.

The 30th annual IDDA Seminar attracted a record food industry crowd. Christison's analysis of the information highway came just as the state Legislature passed legislation to help open the new highway in Wisconsin.

"It's real easy to get on the information highway," she said, "if you've got a cellular telephone, a pager, a computer modem or a satellite dish, you're already on the highway."

The big step that's coming soon, Christison continued, is DBS or Direct Broadcast Satellites, to give people a direct link with everyone in the world.

"Your old telephone is really land-based technology using copper wiring. A lot of the new technoldogy will be wireless, will connect your computer, interactive TV, satellite dish, pager, cellular phone, wireless pocket communicator, video phone or cellular personal digital assistant."

"All this technology sounds great, but what does it mean to the way we do business? In most cases people will be using new technology to do the same old things, in the same way, only faster. What we need to do is to learn how to use new applications to change the way we think," she added.

While people are looking at the new technology, it is being applied with the same rules people always had, she said.

According to Christison, "we have to change the rules of the game. We have to look at new technologies and say, what if we do it this way? Or that? That's our biggest challenged and that's why we're on this information highway."

Christison said she believes the new technology in real terms means "you'll able to do that which you have always wanted to do if you only knew you could do it and you'll be able to do it faster than you ever thought possible."

For instance: Without ever leaving your armchair,

you'll be able to do your banking, take a class or enroll in school, shop, pay your taxes (and get the refund wired to your bank), visit with your neighbors, play games, have a video conference, order and watch movies without ever putting in a cassette, read video newpapers, get an illness diagnosed, attend electronic town meetings, vote by electronics, participate in virtual reality explorations, use a neighborhood network, a global personal E-mail network, order your favorite take-out, use digitally compressed video to change reality and pay bills.

Christison's research shows that satellites will be supplemented with fiber-optic cables with glass strands thinner than human hair. But a single strand of fiber can support 250,000 times as much data as a standard copper telephone wire.

She maintains that the new technologies have to be easy enough for grandma to use, which means that video recorders no longer will flash 12 o'clock.

The total price tag for the highway will be about $500 billion, she added, "and your cut will be about $1,000. That's the good news."

The bad news is that special network switching systems will be developed to handle all the data. And with these new (computer) software systems, government agencies will be able to collect masses of information about you by monitoring your telephone patterns, and credit card transactions — all without actually evesdropping or obtaining permission.

"There also will be computer hackers, thieves, and terrorists of cyberspace who will pilfer thousands of passwords and steal everything from information to your money. And if they are really good at it they won't even leave a vapor trail," Christison warned.

One of the future developments coming to the system are micro-chips about the size of the head of a pin, Christison continued.

This little addition embedded in the label or package of every item in a grocery store, means that each item is scanned as it is put into the cart and the check out completed in six seconds.

Carol Christison

"Today's consumer doesn't have time to do just one thing at a time," she told food business representatives. Time is the currency of the '90s, the cost of something measured in minutes, not dollars. So much so that supermarkets of tommorrow will be known as Service Centers or Service Marts because of one-stop, time-saving conveniences they'll offer.

He's a Classified Expert on Cattle

November 1, 1992

WATERTOWN — There are a lot of people in dairy cattle businesses who know breeders and cattle in fantastic numbers.

But Merle Howard, of Watertown, must be tops in the nation and internationally.

Anyone who follows Holsteins knows there is hardly a sale, show, or major cattle event that Howard hasn't attended.

And to update you on Howard's travels, he has just

returned from the Southern National Show in Missouri and the Western National Show in San Francisco, plus the Wine Country Classic Sale at Santa Rosa, Calif.

He is a walking cattle encyclopedia and quickly tells you that 117 head sold for an average of $6,000 and 102 head went through the Southern National Sale to bring an average of slightly more than $7,000.

Howard's analysis of what's happening in California dairying right now: "If you can't make any money with 500 cows, you put in 700 head, see. That's the way those people operate out there." Compare that with Wisconsin where the average cow herd on a dairy farm is 50 to 60 head, while herds with 100 cows are increasing.

And something like the California dairy trend is occurring in Wisconsin, too, with some smaller numbers, "so if you can't make it on 100 head you go to 150 head. I hate to see the smaller farmer breeder herds go out of business entirely, but we are going to see a decrease along that line and see a predominance of larger herds.

"The owners of such farms are not going to run the farms, they are going to concentrate on the dairy operation, while crop production and field work will go to others," he said.

Howard got his start in Whitewater, where he was born 75 years ago next February, and grew up on the family dairy farm east of the city.

This farm was settled in the late 1800s. His sister, Mary, and his brother-in-law, Fred Mason, have the family farm today.

In the late 1930s he went to the UW-Madison and worked his way through school in the university dairy, living with seven other students in a room over the dairy barn.

A job compiling Guernsey records kept him in a few dollars, but it was living on the reduced prices of day-old products at the 20th Century Market that kept expenses down for the student dairy workers.

He was such a faithful customer of day-old products that clerk Verona Stephenhagen, Arlington, gathered the reduced price products for him each day.

Merle and Verona later were married, and Merle completed his education in dairy husbandry and agronomy in 1942. He also was a member of the UW dairy cattle judging team, through which he met Gene Meyer of Iowa State and Jim Cavanaugh of Purdue, who were major judging competitors.

In 1958, he was named a Holstein Association classifier and for nine years worked to classify herds throughout the country.

This job is where he got to hundreds of farms and classified cattle by giving each animal a numerical rating, which translated into standards like "very good or excellent."

At times he was on a farm two or three days to complete the job and traveled to dairy farms from New York to California and Washington.

In 1968, he joined International Holstein Sales and Service with Al Piper at Lake Mills. He handled export sales, primarily to Central and South America.

Howard is now in livestock insurance with the American Livestock Insurance, Harding and Harding.

High-Tech Cows Bound for Lake Mills

February 23, 1992

LAKE MILLS — A specialized California business provided a February surprise for the Wisconsin economy.

Al Piper, a Lake Mills area business development enthusiast, made it possible for a California firm, named

Bova-Max, to move here when he sold his 65-acre farm to Ralph Sikkema, of Oakdale, Calif., a veterinarian and leading California dairyman.

The farm he purchased is on Manske Road, off Highway 89, about a mile north of Lake Mills and will be leased to Bova-Max.

Al Piper

Sikkema is treasurer of Bova-Max and Robert Muirbrook, Hanover, Calif., is president.

Bova-Max is a service agency for dairymen who have exceptional cows, but are unable to get them bred to produce quality calves to realize the value of their investment.

Muirbrook describes the firm's service as advanced bovine reproductive technology — a generation beyond conventional embryo transfer. It is being set up, he said, to serve dairy cattle breeders in the Midwest and the East.

There are three other phases planned for the Lake Mills site:

1. To be home to an "elite herd," a selection of cows from Sikkema's Star-Holstein herd and other cows with high pedigrees, production and superior type Holsteins to be purchased in this area. The herd should be assembled by the end of the year.

This 30-cow elite herd will be shaped to have "bull mothers" — the cows of such quality that they will be bred for production of young sires sought by artificial insemination firms for testing.

2. An embryo collection business for embryo exports and domestic demand. Sikkema said embryo exports cannot be approved from California because of blue tongue disease. Wisconsin is free of the disease.

3. There also will be a 100-cow milking herd on the farm to provide the cattle needed for embryo transfers. Bova-Max apparently is the first dairy-related California firm to move to Wisconsin. It should start operations here in 60 or 90 days. Interviews with prospective employees have started.

To get calves from cows with breeding problems, the cows will have to be be brought to Lake Mills for the procedure.

In a research and development program at Bova-Max headquarters in Hanford, Calif., techniques are used to take the immature eggs from a problem cow and then when the eggs are mature, use in vitro fertilization to get the embryo needed to develop the calf in a recipient cow.

Robert Muirbrook, 49, was born on a Utah dairy farm and was an agribusiness graduate of Brigham Young University.

For a number of years he managed dairy herds of 600 cows in Arizona.

Before he joined Bova-Max, he managed the Holstein herd for the California unit of Granada Bio-Sciences and worked on cloning research.

He knows the Lake Mills area because he was a frequent visitor to Piper's Farm, where Granada's Wisconsin branch was in operation. Both the California and Wisconsin units were recently closed.

Muirbrook said that the Bova-Max experience in getting calves from the eggs of problem cows has had such a high number of successes that they now can offer the service for dairy breeders. Such a service costs about $2,000, but the total includes the cost for the recipient cow, which has a value of $1,200. Cost of the service itself varies from $500 to $800.

Sikkema, who is in his early 50s, is a native of Holland. He received his veterinary medicine degree at the University of California-Davis and has two 1,000-cow milking herds.

The one at Oakdale is producing nine gallons of milk per cow per day and the herd has an annual average production of 23,400 pounds. Oakdale is about 11 miles from Modesto in Central California directly east of San Francisco.

He has his second herd at Raisin City, but it has not been on test long enough to establish a production record.

Bova-Max was formed by 12 investors a year ago and Sikkema said all the shares are taken, "at this time."

In addition to being in an area free of blue tongue disease, there were other reasons Lake Mills was chosen for breeding service and the elite herd.

Muirbrook said the location is close to World Dairy Expo in Madison and the proposed home of the World Dairy Center.

Bova-Max also will have a joint research project in breeding technology at the UW-Madison, through Professor Jack Rutledge, chairman of the department of Meat and Animal Science.

Pipers' Operation a Breed Apart

February 7, 1993

LAKE MILLS — When Al Piper of Lake Mills recounts his days in the Holstein cattle business it adds up to more than 50 years.

And he did it in the pioneer days of the cattle export business. He was an early leader in helping build the trade balance in Wisconsin's favor.

Piper, 71, opened strong cattle markets from Russia, Iran and Turkey in the East to all the major countries of Central and South America in the West.

In addition he supplied a vigorous domestic Holstein cattle market in the U.S., particularly in the South.

The Piper Brothers — Alvin, Arden, Carl, Milton and George — developed the family farm south of Watertown into an 800-acre operation.

The Piper Brothers' progression in the business world started with the family farm. They developed the business into a stud service that included more than 50 sires for the artificial insemination market.

Add to that the Piper Brothers Dairy that Al built to serve bottled milk demand in the Jefferson and Dodge County areas.

The brothers also formed the Piper Brothers Holstein Sales business serving Wisconsin breeders and those throughout the nation.

The stud service was purchased by Carnation Breeding and now is owned by Landmark Genetics.

Al and Marian, the Jefferson County home agent, were married and moved to Lake Mills in 1969.

Marian was as well-known as her famous husband because she was the clerk of all of Al's sales. Their daughter, Mary, has all the business skills of her mother, father, uncles and grandfather and formed the Lake Mills Travel Co.

Mary knows the Holstein business, too, and has organized a special spring trip to Arizona for those who want to visit farms and see the Milwaukee Brewers in a spring training game.

Al formed the International Holstein Sales and Service firm in the 1970s.

He recounts an especially good sale when he shipped four planeloads of bulls to Greece.

And he worked with a relative of the shah of Iran to build up the cattle numbers in the developing oil country. But later, his prime market became Libya, which was his most profitable sales area.

In filling the demand for cattle in Peru, he shipped more

than 1,000 head, some as recently as 1990 and 1991.

In 1979, he and Bob Strickler started a Holstein sales spectacular in the auditorium of the MGM hotel in Reno. Buyers from throughout the nation flocked to the sale in such numbers, Al said, that they practically had to close down the gambling machines at the hotel.

After all, buyers made a safe bet on Holsteins and the first sale total topped $1 million.

Four sales, one every other year, were held. Al said when he started in cattle sales in the 1940s, prices were nothing like they are now because there were more cattle available.

He said those early sales brought bids of $500 to $600 a head. In contrast, to get quality animals today buyers are bidding $2,500 to $5,000 a head.

Al is a cattle sales historian and has sales catalogs dating to 1916 when John Hetz, Fort Atkinson, was the premier breeder.

Al is out of the cattle sales business today, but he has teamed up with Marlowe Nelson, Westby. Guess what business he is in now?

He and Marlowe are selling semen to breeders who want high protein bulls. Their leased Holstein bull has the highest protein percentage of the breed and the highest type rating of the breed.

Huber's Commute, Career Grow

December 15, 1991

ST. DAVIDS, PA. — When Carl J. Huber first went to work in a cheese factory, the total one-way commute was 13 steps.

That's when he, as a young boy, walked down the stairs from his parents' home to the cheesemaking room at the

Lakeshore Cheese Factory, operated by his father, Alois, in Dodge County, three miles out of Beaver Dam.

But things change and Huber now has a 50 mile commute — one way — from his home in Allentown, Pa., to St. Davids to get to his office at the corporate headquarters of M.E. Franks Inc. The firm also has a smaller Minneapolis office.

St. Davids is on Highway 30, along the Main Line west of Philadelphia.

Huber, now 60, is one of the market managers at the M.E. Franks firm, the largest domestic and international operation of its kind — providing ingredients and services to food and feed industries.

The firm is still run by its founder, Melvin E. (Mel) Franks, president and chief executive officer, who was honored this month as the Distinguished Citizen of Agriculture.

The Franks organization is recognized in the dairy trade as one of the most active and well-known dairy marketing firms in the United States.

Huber joined the corporation in 1985. His position involves the purchasing of dry milks and whey products for customers at home and abroad, who use them in manufactured foods and a variety of other products.

Huber's career plan was to regularly broaden his professional expertise in all areas of dairy products and management.

As a first generation Switzer, Huber got a sound start early at the family cheese factory. He later graduated from Wayland Academy in Beaver Dam.

He then studied the dairy industry studies under UW-Madison Prof. Walter Price, a noted cheese expert, and earned his degree in 1953.

Huber had many student jobs on UW farms and in laboratories, including those in the dairy department in Hiram Smith Hall, where it was located before Babcock

Carl Huber

Hall was completed.

He also was in the communications phase of student life when he was advertising manager and, later, business manager of the Wisconsin Country Magazine, a student publication on the agricultural campus.

When he graduated, he was commissioned as an Army officer through ROTC and served two years of active duty. He married Carol Wolf, a grade school teacher from Dodge County, and they have five children. They now have five grandsons.

After his discharge from the Army, he started a 13-year career with Lake to Lake Dairy Cooperative. First in milk quality work before being named assistant plant manager at the cooperative's Kiel plant and its cheese operation.

It was during this term at Kiel that he became known as one of the cheese greats in Wisconsin when, in 1966, he was presented with the first World Champion Cheese award and accepted the honor as representative of the Kiel plant employees.

Huber was on his way then. His planned career broadened with a move to Denmark, Wis., as plant manager. This is where he learned the technologies of butter and milk powder. The emphasis was on production of quality milk and its products.

After Lake to Lake, he joined Consolidated Badger for seven years, starting at the cooperative's cheese operation in Marshfield.

Then he joined the Laprino Foods operation in Central California, and began working his way East to manage Italian Cheese plants of Cudahy Foods in Kentucky and then moved to Lehigh Valley Farms, a cooperative, at Allentown, until it was purchased by Beatrice of Waukesha.

His time with Lehigh Valley was his learning phase in cultured dairy foods.

Now, as a market manager, Huber is in contact with dairy product and ingredient firms throughout the country. The major change Huber has seen in his 36 years is that dairying has moved into a world market.

"Now that it is a global business we have to balance any excesses in production by being able to market them on a level playing field.

"We are not going to have government with us in dairying forever. You've already seen that," he added.

"So when the government doesn't help us, what do you have to do with all the excess product. You sell it. But then, it has to be competitively priced," Huber explained.

Right now most of the international dairy trade for the U.S. is through the government's Dairy Export Incentive Program. "That's what you have to do. You have to equalize the prices. It is the program that makes us competitive in world markets and helps move excess dairy production out of the country," he said.

Embryo Transfers Growing Business

July 26, 1992

Dr. Robert F. Rowe, the Middleton veterinarian whose research gave dairy cattle breeding one of its technological victories, has expanded his dairy and beef cattle embryo business to Brazil.

A native of Kansasville in Racine County, Rowe's UW-Madison research developed a system for non-surgical removal of cattle embryos with a procedure known as a flush. There were two other competing scientists who developed the flush method of embryo removal, but of the three, the 1976 record established that Rowe was the first.

The breakthrough came while he was a graduate student at UW-Madison working under Dr. O.J. Ginther and specializing in embryo transfers to cows that would carry the calf until it was born.

Rowe received his veterinary science education and doctorate at Michigan State University because Wisconsin at the time had no veterinary school.

Among the highlights of his MSU years was his rank as the top individual in the Collegiate Dairy Cattle Judging Contest and winner of the Kildee Scholarship. He developed his dairy cattle judging skills through Wisconsin 4-H work.

After graduation, Rowe had a number of job offers, but he was determined to return to Wisconsin "because I wanted to work for dairy people I had grown up with."

He started first with a veterinarian in Lodi, and after completing two years as an Air Force officer and base veterinarian in Texas, he returned to Wisconsin to work with a veterinarian in Lake Mills.

Soon after his return he started working with Wisconsin breeders like Howard Voegeli, a Monticello Brown Swiss breeder, to export embryos to South and Central America.

This was before the technique for freezing embryos was perfected and Rowe flushed the embryos from the cow six or seven days after she was bred, checked their quality, and then put them in a thermos-type container, drove to O'Hare airport, and flew the embryos to the farms in Colombia, Equador and Venezuela for transfer to recipient cows.

A short time later, the U.S. Holstein Association hired him to make the same warm thermal transfer of embryos from U.S. cows to surrogate cows in Hungary and Spain.

The system for freezing embryos came in 1983 and 1984. Although the percentage of successful transfers dropped with frozen embryos, it was the most practical method.

So it was natural that Rowe would go into the embryo business to capitalize on the techniques he perfected over 17 years.

"I love the people I work for. They are the best people in the world, whether it is in Moscow or Brazil or whatever country demands . . . services. Remember, I grew up on a dairy farm and that is something special in this business," he says.

But he has to have a supply of embryos, which in the case of his Brazilian account means he must secure 500 quality embryos to meet the $250,000 order.

His embryo sources are dairy farmers in Wisconsin, whom he enjoys working with because "I can pay back the people who are leaders in cattle breeding."

His biggest joy comes when he returns to the farm of his embryo provider and turns over checks ranging from $2,000 to $6,000 — "money that really helps, especially in times like these dairy farmers have had."

What's more, Rowe's Brazilian contact wants to expand. He not only is using the embryos for his own herd, but for other Brazilian cattle breeders — beef as well as dairy — who have learned of Rowe's successes in embryo tranfers.

In the latest shipment of 126 embryos, the costs ranged from $350 to $500 for each embryo.

The Rowe enterprise developed as a result of contacts made at World Dairy Expo last October. The Brazilians stayed on after the Expo and completed their arrangements and signed a contract.

A month ago, the Brazilians returned to Wisconsin and met the dairy farmers who are the source of the needed embryos. They also visited the farms and became acquainted with the cow families they rely on for quality embryos.

"What the world understands is that the U.S. has the

superior line of cattle genetics in the world. They want to come here and buy equipment they need and (get) management skills as well as an understanding of nutrition, breeding, feeding and crop production," he says.

Popular Artist Inspired by Dairy Farming

November 6, 1994

Bonnie Mohr is a devoted, self-taught artist, who has added culture and a beautiful personal philosophy to dairy agriculture.

For nine years, this gifted artist, a dairy farmer's wife from Glencoe, Minn., 50 miles west of Minneapolis, has been coming to World Dairy Expo to offer visitors her paintings of cows and agricultural scenes.

Expo visitors from throughout the nation and the world have responded by buying thousands of her signed and numbered prints to hang in their homes and offices.

She is such a popular figure at Expo, always at the same show location — just inside the west entrance of the Memorial Coliseum — that getting a full interview with her was a three-year task.

Finally, she granted 20 minutes of her time, but as we talked near her show location, customers kept stopping in to chat — including a draft horse enthusiast. Said Mohr, "I always wanted to paint teams of horses doing fieldwork."

Mohr has built the World Dairy Expo into her largest annual show.

After a slow start in 1985 and 1986, Mohr has probably become one of the most productive artists in the nation for her specialized audience. And thanks to the international flavor of the Expo, Mohr's audience extends far beyond the borders of the United States.

Bonnie Mohr

What they like about Mohr's paintings and prints is that she reflects the spirit of dairy farming, which is how most of her customers choose to make a living.

Furthermore, Mohr's large prints are reasonably priced at $40 to $80 — so many people can enjoy art works in their homes.

Mohr and her husband, John, have two children, Amanda, 4 and Katelyn, 2, and while working the 1994 show in Madison she reminded visitors that within another month or so she would have her third child.

Mohr is painting full time now to keep up with growing demand for her work. She has added to her business volume by including note cards, each with renditions of her famed art works.

To help her paint full time much of the week, she has a woman from the neighborhood come in to be with the children.

Mohr is like many farm wives, getting up at 4 or 5 a.m. each morning to paint, just as she did for 11 years when she and her husband, John, did milking and barn chores together.

John has barn help now.

"I painted things that were natural things for me," she recalled. "I was raised on a dairy farm and had an interest in art and a desire to paint, so when I started I chose the environment I was in — farming and cows."

Her favorite painting is entitled "Under the Autumn Oaks," a quiet scene with a few cows grazing and some lying down in a pasture under the dominating oaks.

Customers have selected a splendid painting, "Daybreak," as the one they find most popular. It shows one standing cow and a portion of the home herd lying quietly on a July morning, "just waiting for someone to come and get them so they can go to the barn to be milked," said Mohr.

The 12-inch by 24-inch print, is probably in more homes than any of her other works. Mohr said her inspiration came from her own herd one morning.

When she went out to get the cows that day, she thought, "Oh, Wow. What a painting!" She hurried back to the house to get her camera, and used the photo as a starting point for the picture.

Mohr said she uses her own experiences on the farm as inspiration for most of her work.

"Each one is a scene of what in my life I totally believe in and want to treasure," she said.

Graham's Dream Unfolds: The World Dairy Center

October 4, 1992

Last week, William T. Graham, an inspiration to agriculture at home and abroad, has caused as much excitement as farmers and business people ever experienced in Wisconsin.

This came with the selection of his site and plan for the World Dairy Center and the new headquarters of the Wisconsin Department of Agriculture, Trade and Consumer Protection.

Suddenly all of agriculture was endowed with a new state and international status that is fitting for the new century.

Even Graham was excited.

At a reception to celebrate this giant step for the men

and women of agriculture and its associated businesses, he greeted hundreds, many of whom will be key players in the World Agriculture Center and International Chamber of Agriculture.

It was also a salute to Graham, who is a dedicated patron of agriculture in its new age.

They came to recognize his vision and planning that ultimately will carry out his dream.

But as a true leader, Graham insisted he had not done the job, but simply was caught up in the movement that happened to express his philosophy "to improve the quality of life and the environment of the world through agriculture."

But how does this successful 68-year-old businessman get to express his dream through agriculture? It's simple.

His roots are in agriculture in Georgia — Rogers, Ga. Does that ring a bell? Remember he founded the W.T. Rogers Co. in Madison and never forgot what agriculture and the community of Rogers in the Depression of the 1930s did for him.

Recently he sold that business to have the resources — in millions of dollars — that would help new generations to move confidently and with reward through the early 21st century.

Graham said he believes new technologies to help producers and marketers are swirling around in someone's head or sitting on shelves, available, but unused.

It will be these technologies he wants to tap.

He didn't have those back in Rogers, where his grandfather lost most of his land and property during the Depression. His family had to move so his mother could take a job as a bookkeeper in a general store at $5 a week.

His father worked as a cotton buyer and when business was slow he was with the Work Projects Administration (WPA), created to provide jobs during the Depression.

Graham was 11 years old when he opened a store of his own in the crossroads of Comer, selling tobacco products he could buy at a 25 percent discount, which was possible because his mother worked at the store.

The enterprise grew and Graham offered a full line of products. He was so involved his father took over the store until it burned about 20 years later.

While in Clayton High School he was a 4-H club member and milked three cows, in addition to raising chickens and hogs, while caring for two large gardens.

William T. Graham

As a newspaper carrier for the Atlanta Constitution he took eggs and other products in exchange for the paper. But he started to see the world when he won a paper sales contest offering a trip to Miami, Fla.

Later with his business success in Madison, after World War II and graduation from the UW-Madison, he kept his hand in agriculture for 35 years through various partnership arrangements in feeding beef, dairies in Texas, and vegetable farming. He also has farms at New Glarus, Jefferson, and recently purchased another farm at Albion, where he wants to put in natural grasses and let the land revert to wetlands.

Graham knows agriculture. Now it's his life that he will share with the world as his concept radiates from Madison to farmers and businesses throughout the world.

Milk Hauler Stays Close to Farm Even on Vacation

May 10, 1992

VERONA — Most people call Lyle Wealti a milk hauler and that's accurate.

He spends 10 to 11 hours a day driving two routes and 140 miles every day of every week — except for the two weeks in late winter when its a trucker's time to take time off and travel.

The only one who drives his $130,000 tank truck and a load of 49,000 pounds of milk when Wealti is gone is his son.

In fact, it is his son who tells Wealti when he can take his vacation.

But for about 20 years Wealti drove his truck every day in every week of every year. There were no vacations then. When he was able to take time off, his lifetime goal was to visit Hawaii and he has been there three times. He and his wife, Ann, cruised the Caribbean and a dozen ports of call, he has relaxed in Mexico at least four times and on his latest trip he set out alone to learn about California's growing dairy industry.

"When I travel," he said, "I always have to be tied in with some farming. Then I feel secure. I really like farming."

Wealti, 55, also is a trucking company owner, a eucher player and team sponsor, an architect when it comes to remodeling farm homes, an ex-Marine sergeant, a one-time operator of farms and holds title as a master machinist, which he earned at the Chicago Institute of Technology under the GI Bill.

But he was a country man. When he went to work for

General Motors in Janesville, it didn't suit him and he soon left to resume his favorite job as a milk hauler.

He had to take part-time work helping farmers with their haying, do custom combining work and help out with rural mail delivery before Christmas to make ends meet after buying his own milk route.

Wealti also is a stimulating conversationalist because he is a humorist and one of those gifted persons with instant recall of names and events.

All his trips are documented with excellent photos and a story.

Wealti was born on a family dairy farm at Basco and attended school at Belleville.

He didn't finish high school until he was in the Marine Corps, but he was a worker from the start. He wasn't quite 16 years old when he started driving a truck to pick up milk cans on farms. The truck was owned by Wealti's mother and himself.

But he was strong enough to handle the route. This was when milk was still transported in cans and he could reach into the water cooler and lift two full cans (160-plus pounds) and load them on to the truck.

The California trip this year to see the dairies was his first trip there since he served in the Marine Corps — 1954 to 1957.

"Yeah, I had to go out and see the dairies," he said. Because California's milk production challenges Wisconsin's rank as America's Dairyland, Wisconsin dairy people are interested in its farms.

Wealti got to places at dairies that few people get to see. On a farm with 2,500 cows, he stood in a milking parlor where four workers milked 56 cows at one time and did it for a full shift. Other workers come in to continue the round-the-clock milking.

The stalls had automatic take-offs, which means the milking unit is dropped as each cow finishes milking.

"What blew my mind is that the dairies there got $3 a hundredweight less for their milk than we get here in Dane County . . . " he said.

"When I left on the trip farmers here were getting $13 for their milk and the California dairy operator said he was getting $9 a hundredweight for his surplus milk and between $10.50 and $11 for the base milk. The price averaged out at just under $10 per hundredweight.

"And to get that base established to get the higher price, the dairies had to pay $2,500 per cow. But nobody was buying the base and everyone was milking at the surplus price of $9," Wealti said. And, they do not receive a premium for high quality milk.

"But the only premium, if you want to call it that, is to keep producing milk with a somatic cell count below 500,000. If that count goes higher, the dairy is cut off from the milk market and the milk can only be sold to cheese plants. Milk with traces of antibiotics can only be used in large calf feeding operations."

Chapter 5

Beef and Dairy Beef

Demand Passes Feeder Cattle Supply

February 25, 1990

Charles Schrader, a Stitzer farmer with a family partnership in a cow-calf operation, has a real challenge off the farm.

As a cow-calf operator, Schrader maintains a 90-cow beef herd and makes his income from the sale of 6- or 7-month-old calves or holds them over for a year to be sold as yearling feeders.

"The next feeders I will sell haven't been born yet. So the calves born this spring will be sold in the fall of 1991," he said.

Feeder cattle are the lightweight steers and heifers raised by breeders and sold to cattle feeders who put them in feedlots to carry them to market weights of 1,150 to 1,300 pounds.

Schrader, 60, who started with Herefords back in 1963, now has Herefords and Simmental crossbreds. He bought his father's farm in 1960 after he got out of the Navy and continued a dairy operation for three years.

Schrader is president of the Southwest Wisconsin Feeder Cattle Cooperative, which for most of its 32 years has been the price-setter for feeder cattle sale in October.

But like everthing else in agriculture, Schrader said the beef industry is changing. His job now is to get more feeder cattle numbers to keep up with buyer demand.

There are fewer and fewer southwest Wisconsin farmers with small beef herds of the kind that was so dominant when the cooperative started its sales in the late 1950s.

Schrader fears that, unless the cooperative can get the number of feeder cattle buyers want, the cooperative will lose their big buyers who run feedlots in Illinois.

Cooperative members voted with mixed feelings to include Holstein steers in the auction schedule. It was a tough decision for many dedicated beef breeders.

Charles Schrader

Some members even indicated that if the organization included Holstein feeders they no longer would consign beef feeders to cooperative auction sales.

But other cooperative members said they were surprised how many buyers said they wanted Holstein feeders. The southwest feeder cattle group is the last unit in the state to include a Holstein feeder cattle sale.

"We are going to have a Holstein feeder in October. We are getting a lot of cooperation for this sale from Dick Klas, manager of the Midwest Sale Barn at Fennimore. He is going to help us get the number of feeders we need," Schrader said.

By holding a separate Holstein steer sale, the cooperative hopes to keep from losing beef breeders who do not want Holstein feeder sales.

Schrader's partnership is with his son, David, and their working partners, Joann, the senior Schrader's wife, and Joan, David's wife.

"My daughter-in-law and I are among only a few farm women who don't have a job off the farm. We devote all our time to our husbands and the farm," Joann said.

She's proud of her farm role in which she does all the hay baling. The abundance of hay on the 485-acre farm and a rented 80-acre farm permits the Schraders to carryover calves so they can be sold as yearlings at 800 or

850 pounds.

The Schraders use a throw-baler which sends the bale flying from the baler to the wagon. Schrader said Joann probably is the best shot there is since she doesn't miss many times in harvesting 25,000 bales a year.

When Joann isn't baling, she's fully engaged in chores with the herd of 90 sows. When the litters are born, she cuts the teeth and tails and later does all the castrating. Most of the pigs are marketed as slaughter hogs.

Joan keeps track of the vaccinations, and when it is time to give the shots in a major herd health system.

The Schraders also have a daughter and two other sons.

Cowboy's Tales Keep 'Em Laughing

October 29, 1989

BLANCHARDVILLE — In the rolling hills along the Lafayette-Iowa County line where, the Herefords roam, there's a cowboy who has the makings of Wisconsin's own Will Rogers.

The cowpoke from Gant Road along the banks of the Yellowstone River is Jerry Gant, 52, a Hereford breeder who can make people laugh, and as a result, feel a little better about themselves and things around them.

Not everyone can carry off the humor and manner of a Will Rogers, but Gant does a pretty good job of it just by letting people feel good with his good, clean humor.

That's an art in itself these days.

Gant is a lot like the famed Oklahoma humorist of the 1920s and 1930s who put laughter and light in the lives of people during the dark days of the Depression.

These were the folks who gathered around the old Majestic radio on Sunday nights.

Gant's appearances before public audiences are rare since he only has spoken to Eastern Star groups of which he is an officer.

But in his debute over at Platteville a while back Gant had his audience laughing for nearly 40 minutes.

Anyone who can pack in 52 bursts of laughter and lingering chuckles in 40 minutes has a touch of a master of laughter.

The thing about this Rogers image is that Gant has a little twang in his voice that only can come when a cowboy tells of his own experiences in his own way.

Right away you might say Gant only talks of the farm and the cattle. But that's not it at all. His little stack of 3-by-5 cards has jokes or thoughts about everything from Belmont to Chicago.

The genius of Gant is that he creates such vivid descriptions of stories on the farm that he carries his audience as if they were right with him. And when that red-eyed cow is breathing on Gant's back pocket as he races across that river bottom pasture, you get into the story so much you are panting as much as Gant in his dash for the one old tree with a single branch where he puts all his energy into a dramatic leap.

Oh, just let Gant tell it. You won't stop laughing for a spell.

His audiences tend to divide themselves. Women get into the spirit quickly and lead the laughter. But not much later the men are having as much fun as anyone.

Gant is at his best telling stories on himself, like the time people observed that he and his brother, Karl, are such extreme opposites that one person asked their mother if there was a possibility Jerry was adopted.

"Well, yes. Twice" she replied."And they brought him back both times."

Jerry Gant

He also has a joke or two from his cards that don't bring immediate response, but like a story having a delayed humor fuse the crowd suddenly catches on. He gets a much better response than if they had caught on immediately.

Jerry and his brother farm on 1,700 acres of land in Lafayette County. They do it all without reliance on government programs even though there would have been direct payments totalling many thousands of dollars last year to cover drought assistance. But that's not the course of action they chose.

Gant's love of humor has always been part his life and one of his added talents was acting in Blanchardville High School.

"I guess I was considered more of a bad actor than anything," he laughed.

But he joined the Masons, and had a chance to get in front of people. This made him a little bit bolder as he took the opportunity to tell his stories.

"It was just kind of a happening, I guess," Gant said.

He was elected to an office in the Eastern Star and with all the officer reports and responses he had to make he decided to be different.

"I started to tell a story to get them to listen and then tried to leave them with something. Pretty soon one story led to another and another." He found his audiences waited for the stories and that's just how this natural humorist got going.

"Now I just enjoy doing it and I have more fun than the audience does," he said.

Thomases' Farm a Group Effort

March 21, 1993

If you regard a family farm as something special in today's agriculture, the farm of Deane and Nancy Thomas, on Highway 18 west of Cobb, is a masterpiece.

The 586-acre beef and hog farm settled 104 years ago by Deane's grandfather, A.E. Thomas, today is a happy, fulfilling place for the Thomases and their farming partner son, Jeff, and daughter-in-law, Tara, both UW-Madison graduates.

Deane, 61, a UW-Madison Farm Short Course graduate, handles the big beef herd and Jeff manages the swine herd.

They have another son, Steven, and a daughter-in-law, Janelle (Keller), also UW agriculture graduates, who live in Sun Prairie, where Steven is based for his work as a marketing agent for the Monsanto Agricultural Co. When the work at the home farm gets rolling, Steve and Janelle will join the family to lend a hand.

The Thomas farm in Iowa County showcases state and national efforts to improve livestock farming and was best described in 1992, when Deane and Nancy received honorary recognition from UW-Madison College of Agricultural and Life Sciences.

The citation described them as "true partners of farm and family life; tireless cooperators and promoters of livestock farming and meat marketing; and ardent supporters of livestock educational activities."

Nancy is co-chairwoman of the beef cook-off at the World Beef Show in April at Madison, which brought in 134 entries, testing 274 recipes by candidates from 24 states and Canadian provinces.

Furthermore, the farm, called DNT Farms, attracts the attention of passing motorists with Nancy's message signs.

The brief messages have become an institution. Some provoke response as Nancy found out when she suggested passers-by "honk if you have spring fever" and drivers honked for as long as they could read the sign.

An Oshkosh family let Nancy know that when they leave Madison for the trips west, everyone in the car starts guessing what message Nancy will have for them when they pass the Thomas farm. Another driver called after reading a religous thought and told Nancy it was "just what I needed today."

Nancy is a full-time farm wife because she always believed that the husband, wife and family should work together to make it a true family farm. This concept of the family farm traces to Monticello where she grew up as Nancy Burgy and where her parents raised Angus cattle and Percheron horses.

"On the Thomas farm we work together and we are proud of our labor," she said.

Deane fills his role in beef organizations, which include president of the Wisconsin Cattlemen's Association, a director of the National Cattlemen's Assn., member of the National Beef Promotion and Research Board in Washington, and sales manager and director of Southwest Wisconsin Feeder Cattle Cooperative.

Deane and Nancy met when she was county home agent in Grant County, which she was assigned after her graduation from the UW-Madison.

Meat Industry Inducts Three into Hall of Fame

February 13, 1994

Stories of the leaders who built Wisconsin's famed meat industry were recalled last week at the annual meeting of the Wisconsin Livestock and Meat Council at Lake Mills.

Three honorees were also inducted into the Wisconsin Meat Industry Hall of Fame, which opened in 1993 as the inspiration of Emeritus Professor Robert W. Bray.

Bray grew up on a farm near Dodgeville and devoted his professional life to the University of Wisconsin.

When Bray himself was nominated for the hall a year ago, he refused induction. But when the nomination was submitted a second time and recommended by the committee, he proudly accepted.

Bray organized UW's meat science program and became the first to hold a doctoral degree in meat science at the university. His career included 18 years as associate dean and director of the College of Agricultural and Life Sciences.

His hall of fame honor came for his meat science educational work and his research leadership.

About 2,500 students who took courses under Bray were undergraduates and members of the farm and industry short course. More than 30 students gained advanced degrees with Bray as their adviser, and today they hold leadership roles around the world.

Hall inductee Milo C. Jones, founder of Jones Dairy Farm, built a family firm that is recognized and respected throughout the United States.

The Jones family moved from Vermont to Fort Atkinson in 1858 and served agriculture as a dairy, grain and stock-raising enterprise. Milo C. Jones, born in 1849, joined his father as a dairyman and made cheese the primary family product.

Milo C. Jones was only 35 years old when he was stricken with arthritis and was confined to a wheelchair.

He decided to make pork sausage commercially, using the recipe his mother developed in Vermont. The sausage was made in a room adjoining the cheese room.

To market the sausage, Milo used direct advertising and placed ads in national magazines in 1903. His marketing innovations brought success to Jones Dairy Farm.

Milo died in 1919 and was followed as president by his daughter Mary. Jones family members have continued as top officers of the family business, including Alan, Edward and now Milo C. Jones, Alan's son, who has been president since 1983.

Inductee Fritz Bernegger came to Wisconsin from Austria in 1926 after being trained as a professional butcher.

He planned to return to his homeland in two years, but his life changed when he entered the meat business.

His success story started when he moved to New London in 1928 and got a job "to make good sausage."

He struggled as a sausage manufacturer, but his hard work paid off and his business expanded to three Quality Meat Markets.

He kept the quality name and the quality products as he opened a packing company and had six people working in 1947 at his stockyard, kill floor and sausage kitchen.

He took on a partner, Billy Schmidt, and supplied all the fresh meat and sausages for his stores. In 1953, the partnership split, with Bernegger keeping the packing plant and Schmidt the retail markets.

Bernegger's business expanded rapidly, gaining federal

inspection in 1969. In 1971, when he saw that it no longer was a family business, he agreed to merge with Consolidated Foods.

But the entrepreneur who got a job to "make good sausage," lived to see his smoked sausage become the No. 1-seller in the nation.

By 1977, the plant that Bernegger built had 900 employees and distributed 3 million pounds of meat each week.

Bernegger died in 1988.

The firm became Hillshire Farm and merged with Sara Lee Corp. and in 1989 became part of Kahn's, an operating unit of Sara Lee.

First World Beef Expo Opens Here

January 17, 1993

Mineral Point's Paul May, an Angus beef breeder and beef enthusiast for 40 years, is spending more and more time in Madison these days.

And it's worth every minute and hour he devotes to the 1993 April surprise he and other beef leaders have for Madison, the state, the nation and the world.

May, who was born and raised on the 296-acre Fernvale Farm, 2 miles south of Mineral Point, is executive chairman of the new World Beef Expo to open in Madison in three months.

This show time and calf sale production is planned to attract U.S. and world beef breeders and cattle feeders to the new technology, the show ring and the sales ring for the birth of a new agricultural event that could match the fabulously successful World Dairy Expo.

The Dairy Expo staff under Tom McKittrick, has been

contracted to manage the annual event at the Dane County Exposition Center.

Furthermore, May, who always has been an innovator when it comes to promoting the beef profession and industry, has been thinking about this kind of event for more than three years.

May has put most of a lifetime into the beef world and his wife, Lucille, has been very supportive of every program he worked on to help junior exhibitors become a major part of beef cattle events.

In 1953, May acquired his first registered Angus cattle. The operation is carried on today by the Mays' son, Greg, in the Fernvale Farm Partnership.

And now May is looking to the future as beef breeders set their quality, taste and efficiency goals for the 21st century.

He said he had been thinking of this kind of beef industry world show for more than three years and things broke right when the Wisconsin Cattlemen's Association followed up on the proposal.

About the same time, state Agriculture Secretary Alan Tracy and his team of marketers from the state Department of Agriculture, Trade and Consumer Protection proposed to use its overseas contacts to build in a foreign visitor base for the new World Beef Expo. The department did the same 26 years ago for World Dairy Expo.

Within a short time, the department had commitments for 100 foreign beef industry representatives to come to the first beef expo. Several herds of Canadian beef are expected in Madison.

"We were all ready to go ahead with beef expo in 1992, but we didn't have a location. For a time they considered using facilties at State Fair Park in West Allis," May said.

But the Madison site was better and organizers signed a three-year contract starting with April 16-18.

The taste of beef part of the show, May said, will be in the two beef cook-off contests for state and international contestants. There is a $2,000 top prize.

World Beef Expo has 23 members on its board of directors, May said. The board meets every other month and members of its executive committee meet once a month.

World Beef Expo a Dream Come True

April 18, 1993

Paul May, the celebrated beef leader of Mineral Point in Iowa County, is one happy fellow.

The dream that he and others of the Wisconsin Cattlemen's Association had for a World Beef Expo is being realized this weekend at the Dane County Exposition Center in Madison.

May is the recognized "Father of World Beef Expo," and is Expo's executive chairman.

"We have been able to put together cattle shows for 15 beef breeds and feature programs both junior and adult competition, beef breed sales, meaningful education sessions, and a trade show that has filled exhibit space from wall to wall, that is a program worth a visit," May said.

May always has supported youth involvement in Wisconsin beef cattle events and insisted on a strong youth program for the world show.

For three years, May has been working and planning for this world event, but it was tough to find a home for the show.

But he had the support the Wisconsin Cattlemen's Association and then the cooperation of the Wisconsin Department of Agriculture, Trade and Consumer Protection, when state Secretary of Agriculture Alan Tracy and markets administrator James Smith offered to help build world participation by using its foreign visitor service, as it did 26 years ago to create the successful World Dairy Expo.

The beef program prospects also were aided when arrangements were made with the dairy expo staff under Tom McKittrick to manage World Beef Expo.

In addition, May was able to secure a three-year contract to hold the show at the Dane County Exposition Center.

May and his executive committee couldn't have picked a better year for the grand opening since steer and heifer prices this year reached record levels to reward beef breeders and feeders of quality cattle - a major appeal to taste and nutrition-conscious consumers throughout the nation and the world.

Also, the beef future is even brighter as cattle numbers throughout the world are steadily shrinking, thus creating stronger demand and marketing base for the favorite product for consumers everywhere.

Even Wisconsin dairy farmers and their 1.6 million cows have a major part of the beef business and share this beef excitment as dairy beef steers climbed to a price level this spring that returned close to $1,100 for each market-ready dairy steer. There also has been stronger beef price offered for each dairy farmer in the cull cow market.

The strong beef market has enabled hundreds of dairy beef feeders to improve their cash flow despite other price trends in their business.

May, 70, has had a connection with the beef industry all his life. He became a registered Angus breeder in 1953. The strong base the operation has with 80 Angus cows, was founded with his wife, Lucille.

The operation is being carried by the Mays' son, Greg, in the Fernvale Farm partnership.

And now, May is looking to the future as beef breeders set their quality, taste and efficiency goals for the 21st Century, which will create stronger marketing programs for the benefit of the industry and its customers.

He always has been an innovator in beef promotion of quality beef production and his work and leadership was recognized earlier when elected for two terms as a director of the American Angus Association.

What May's record shows is that he has national support to go with the respect he has earned in Iowa County, his home county, and in Wisconsin. Now he will gain world support with the new show.

His 1993 honors include Iowa County Cattleman of the Year and recognition of the Wisconsin Cattlemen's Association as State Cattleman of the Year.

Woman Devoted Lifetime to Cattle

August 4, 1991

MONROE — Frances Rufi is still winning trophies for showing cattle - just for fun.

All of the towering gold and silver trophies that stand like a brilliant, shining forest of precious metals on a table in her Monroe home were for serious competition.

But her latest acquisition was for being one of the "crazies" among the elders who hammed it up for the cattle crowd at this year's Green County Fair.

Rufi, 63, is having fun now as she has had all her life handling cattle, milking and doing farm field work.

And even in semi-retirement, she is responsible for a 239-acre farm at Twin Grove.

But showing cattle was her life.

Furthermore, she is a woman of distinction - a first in the international show ring.

She rose to the occasion in Japan, where women are not allowed to show cattle or be in the ring.

But in 1980 at the Hokkaido Holstein Show, when the grand champion bull was to leave the ring, she was encouraged by cattle entrepreneur Marlowe Nelson, of Westby, to take the champion back to its stall.

Frances has been breaking down barriers for women all her life and was the spirit of the Rufi farm during its greatest years (1930 to 1978), which came with the breeding of fine Holstein cattle.

Frances was and is the wonderment of the cattle fraternity.

She stands all of 4-feet-11 inches, some say, even 5 feet, but in the days when bulls where regularly shown, Frances could handle a bull weighing 2,300-plus pounds with such

Frances Rufi

considerate and understanding authority that the bulls gave her their complete cooperation.

And when she describes the drama of taking a bull or a cow through the ring and winning a championship, you know she walked back to the barns feeling as if she were 7 feet tall.

"I never was afraid of a bull," she said.

"They became used to me. I handled them and walked bulls from the time they were little calves and it was more like a trust that was built up between us," she continued.

"But you always had to keep an eye on them," she cautioned.

Anyone who meets Frances can immediately become friends. She exudes confidence in people and has a memory file that recalls the circumstances of an earlier meeting.

These hundreds of memories are carefully contained in five huge scrapbooks started by her mother, who found every news clipping and tabulation of cattle show results - like those describing the Rufi Farm, with Frances leading the bulls and winning 15 of the 16 grand champion bull awards at the Tri-County Black and White Show while it was in Evansville.

Frances flipped pages of the books and looked up, saying, "when I look at these pages, it all comes back to me."

The Rufi record recalls the farm's success with the famed Homestead Bull of yesteryear.

The Rufis gained breed recognition for having bred and owned the first gold medal sire - King Cascade Ollie Homestead. This status came as a result of the milk production records of King's 28 classified daughters and their dairy character.

Frances has carried on the scrapbook job since her mother, Eva, died.

And Frances said one of her best learning experiences came when she worked her way through the grades at the one-room Greenbush School. She completed her education at Monroe High School and after graduation immediately went to work on the 251-acre home farm in the Town of Sylvester.

She said she kept so busy she never married.

Frances said she always looked forward to showing cattle at several fairs because to her the annual ritual was like a vacation.

Knebel's Career a Prime Choice

January 2, 1994

Early last summer when a select group of five Wisconsin meat processors were honored as the first to enter the state's new Hall of Fame honoring those in the meats profession, Belmont's Clarence Knebel was a unanimous choice.

And earlier, Clarence was chosen to be in the inaugural group in the National Hall of Fame in Pennsylvania.

Knebel, the exceptional resident of Belmont, died Dec. 13 at age 79.

In the days that followed, the many memories of his quality meat processing and business leadership were recalled by customers of the famed Knebel Meat Processing plant, the meat animal producers of southwest Wisconsin, and fellow members of his profession throughout the state and nation.

For 42 years, until their retirement ending in 1989, Clarence and his wife, Lela, left for work and opened the Knebel plant and market doors at 6:30 a.m. and were lucky to be able to close those door at 6:30 p.m. On Wednesdays and Saturdays it was ll p.m. before they got home.

Their biggest business days were when Iowa football fans returned home from a Camp Randall game in Madison and stopped at the plant to stock up on sausage and meats.

Business partners, Clarence and Lela established a unique and wonderful relationship with area customers and meat animal producers.

By knowing both sides of the meat business - the desires of customers and the abilities of farmers to produce quality meat animals, Clarence was able to carry out his business and personal philosophy.

He often recalled that he set the philosophy of his business early on when he decided meats production "has to be the best or we don't do it at all."

He worked closely with the UW-Madison meat authority, Bob Bray, who said Knebel's leadership helped improve the entire meat processing industry.

And what an impact he had, not only in Belmont and the meats business, but throughout the southwest area, state and nation.

In his busy life, Clarence always had time to serve in his state and national meat professional organizations and on boards of directors for state and university organizations and banking and telephone organizations.

Robert F. Thayer, former markets chief of the Wisconsin Deparment of Agriculture told how dedicated Knebel was to enhance the standing of state meat processors through state meat inspection equal to the federal inspection program.

Clarence worked tirelessly in Washington as a lobbyist for state and national meat processor organizations in an effort that successfully convinced Congress to have the federal government establish state meat inspection matching that of the federal program.

Clarence got a start in the profession from his father, who was a butcher in Benton.

The Knebel product line through the years expanded to 72 varieties of sausage and cured meat items.

Also, Clarence was a cooperator in a program to take meat research and marketing for quality pork products to the business and farm production level.

He teamed up with Richard Vilstrup, UW-Extension livestock economist, and with area farmers to launch the Mor-Lean Pork program, one of the first such quality pork programs in the nation.

Clarence helped put a meat cutting course in the Southwest Vocational Technical School and formed the meat arts course in Belmont High School vocational agriculture unit.

The best part of it all is that Knebel plant ownership and its business philosophy has been taken over by his two sons-in-law, Ken Bisarek and Nick Galle, and Nelson NeCollins, a plant employee.

Belmont Man Wins National Honor

September 1, 1991

BELMONT — The most famous partnership of Belmont — Clarence and Lela Knebel — had good reason to celebrate last week.

Knebel, 76, is now a member of the American Association of Meat Processors Hall of Fame.

He was selected in the inaugural year of the Hall of Fame and recognized on the basis of his "exemplary performance in the craft of curing and smoking meat products."

Celebration at the Knebels' home on Wood Street meant that Clarence felt good about this honor because it also reflected on his employees - "wonderful help" who worked with him. It also involved the Belmont community that in the early years of business asked him to open a retail meat market so they, too, could share Knebel's good meats.

Knebel, from the very start in the meat business was a winner because he lived, worked and carried out a basic business and personal philosophy: "I made up my mind that it has got to be the best or we don't do it at all."

In carrying out this philosophy, Knebel started by buying the best cattle and the best hogs. That led to the best meat products, he said. Furthermore, his employees had something to shoot for and "I insisted they would do it and they were happy to do so."

His early skills as a meat processor and industry leader came from his father who was a butcher in Benton.

He also worked on a farm at Depression wages and through a friend got a job driving new cars from the plant in Kenosha to car dealers.

At work one day in 1941 a fellow employee asked Knebel if he had checked his draft number lately. He hadn't but he went back to Lafayette County and learned he had been drafted and left for the Army the next day.

His leadership qualities were quickly recognized and he was commissioned and served in the Army five years. He earned five battle stars in Europe and Africa and a bronze star and was discharged in 1946 as a captain in field artillery.

That also was the year he went into business with a $13,000 loan and bought a building that had been a drugstore on Mound Street in Belmont.

That year he also met Lela, the girl from Mineral Point and a UW-Platteville graduate and elementary teacher in Monroe. They met at a summer school dance in Platteville. They were married in 1947.

This partnership worked and "we've been dancing ever since - sometimes around the kitchen table and sometimes around a meat block," Lela added with laughter.

But for 42 years they left for work at 6:30 a.m. and were lucky to lock the door of Knebel's Meat Processing Plant by 6:30 p.m. On Wednesdays and Saturdays it was 11 p.m.

before the place was closed. It wasn't only a business the Knebels created, but it became an institution for Belmont residents and farmers of the area.

This was a start of new technology in meat handling and Clarence and his workers cut and wrapped meat for a quick, sharp freeze. Some was stored in the locker plant, but home freezers were more convenient. The demand could not be met, so Knebel's temporarily sold home freezers to their customers.

"I was just trying to get ahead," he said.

Once Clarence and Lela started, the plant on Mound Street expanded again and again. The north wall of the meat market shop became an awards gallery and a record of meat business honors and achievements.

Their family expanded, too, with four daughters. Even with all the business pressures, to Clarence and Lela, the home was supreme.

Knebel's leadership extended to numerous community, area, state and national organizations. He always had time to serve people, others of his profession,serving on state and university, banking and telephone board of directors.

He worked closely with the UW-Madison meat authority Bob Bray, whom Knebel said had an unbelievable impact on the meat industry.

At the first meeting of the Wisconsin Association of Meat Processors, Knebel and Bray organized the first state meat products show for its members.

Lela said and outstanding quality of the state meat processor family and that is a willingness to share secrets of production to help members understand processing methods, with older ones helping younger members.

Together with Richard Vilstrup and with farmers the Mor-Lean name was selected as the brand name for all pork products in one of the first farmer-meat business programs to emphasize the lean qualities of pork by selection of only No. 1 hogs.

His product line expanded to 70 varieties of sausage and cured meat items.

In two of the unique programs in the nation, Knebel established the All-Around Butcher-Meat Cutting and Meat Inspector programs at Southwest Vocational Technical School, Fennimore, and a Meat Arts Course in vocational agriculture at Belmont High School. He did the Belmont teaching in his plant.

Knebel retired in 1989, but both he and Lela get to the plant frequently just to see how things are going.

Plant ownership is being taken over by two sons-in-law, Ken Bisarek, husband of their daughter, Kristine, and Nick Galle, husband of their daughter, Joy; and by Nelson NeCollins, a plant employee. Bisarek, who has a meat plant at Bloomington, has taken over Knebel's state meat processors directorship.

Trader Offers Strategy to Get Beef Growers Through Tough Times

January 21, 1996

CASTLE ROCK — A veteran beef and grain commodity trader and adviser to farmers is suggesting a difficult strategy for price-stressed beef producers who want to stay in the business.

James "Sig" Sigmon, vice president of the Securities Corporation of Iowa, in Cedar Rapids, with more than 25 years in grain and livestock trading, is well known as a regular analyst on public television's popular "Market to Market" program shown across Wisconsin on Sundays at 9 a.m.

At the 38th annual meeting of the Southwest Wisconsin Beef Cattle Cooperative, he told the crowd of more than 100, that the "bad news is that there will be liquidations of the beef cow herds."

Sigmon's talk to farmers also was a key to consumers who haven't been aware at times of the struggling beef industry and how it is affected by the sharp reduction in corn and feed grain stocks in the past year.

He added: "Frankly, we have to liquidate. We do not have enough grain for feed and unfortunately beef herds are going to have to take the cut," Sigmon said.

But even with the approaching liquidation over the next 18 months, Sigmon said, there is hope for cattle prices "to turn the corner in 1997," so those sticking with the cattle business "can earn the big bucks coming in beef."

Sigmon's recommendation to beef producers, including a growing number of dairy beef steers, is short and direct with the use of key management techniques, like protecting your feed prices with forward pricing and especially "being able to pay the banker, who then will agree to lend you more money."

Also, beef producers will have to fully recognize that their future is geared to the export market to fill the demand for beef that is increasing rapidly in Southeast Asia.

Sigmon told producers "we are in a world market and your world is bigger than Southwest Wisconsin. Every time export demand grows, it means we have more mouths to feed."

The export market has become a strong one for the U.S., he continued, because the Corn Belt is the only place in the world that can put the kind of finish on beef cuts, that the growing Asian market demands.

He swelled the crowd's interest by telling them how excited he was about the future of agriculture -- even with present problems.

"I really am excited," he said, "because we are going into the most exciting period in agriculture that you folks will ever see -- and you are going to be part of this world and national development."

Sigmon said there is confidence in agriculture because "you can do whatever you can do best. And right now you are producing better than anyone else in the world."

He came down sharply on talk of a grain embargo being imposed by the United States to cut shipments aboard.

The embargo has failed in the past and it should never be used again, he said.

Chapter 6

Dairy Events

Pink Prepares for Yet Another Challenge

June 7, 1990

LANCASTER — William Pink, a practical historian and recorder of farm and rural community events, is looking for a chance to make history Sunday.

Pink, 61, is host of Grant County's third annual Dairy Breakfast on the Farm to be held from 7 a.m. to noon. The Pink Farm is on Badger Road, two miles west of Highway K, which runs between Lancaster and Mount Ida. Signs will point the way. The breakfast is sponsored by the Lancaster FFA Alumni.

Pink believes there is enough public interest in Grant County and the surrounding area to set a state record for dairy breakfast attendance. Even the Green Bay Packers are cooperating with the event by bringing in Ron Holstrom, offensive lineman, to meet people and give autographs.

As a dairy farmer, Pink has faced challenges all his life. His reaction to the suggestion that his farm play host to the state's biggest June Dairy Month event is typical of Pink's philosophy:

"There is no reason we can't, is there?" This view must be his working philosophy because the Pink family has been farming since 1848, the year Wisconsin became a state.

Pink can trace the Michael Pink family of Germany to their arrival in America 142 years ago at New Orleans. The family took a boat up the Mississippi to Galena, Ill. His great-great grandfather then went north and found land for their first farm near Tennyson. Later the family moved further north in Grant County to Hurricane.

When his father, Albert, started farming on his own it was on a rented farm at Livingston, where Pink was born, and then moved to a farm near Cobb.

In 1936, the family purchased the 230-acre farm where Pink lives today.

When his father died in 1954, Pink took over the operation because he was the oldest.

His farm home contains many family heirlooms. He still has the 100-year-old mantel clock that was his grandfather's wedding present. Pink has the clock strike the hour to demonstrate its pure, clear tone.

The size of the farm has steadily increased to 630 acres. It always has been a dairy farm. Pink and his brother, Clement, farm together and their two farms total 1,100 acres.

There are 40 cows in the barn today, but Pink can vividly tell you of the time when there were 20 cows and all had to be milked by hand.

"There were four of us and we each milked five cows — the same five each morning and night," he said. "The cows I remember were Dynamite and Ol' Red Cow," he added.

That herd remained at 20 cows until the milking machine came along and the technology was so wonderful that the milking herd was increased to 38 head.

But Pink had some regrets and reservations about the end of hand milking, "because to me, sitting down alongside a cow and milking her by hand is relaxation. That's real relaxation.

"We always used to have the lantern in the barn because we didn't have electricity on the farm," he recalled. In 1943, the Pinks' purchased their first purebred Holstein heifer, Locustdale Bessie Homestead, from the Melfred Bohe herd for $25. He pulls out a picture of the farm's second purebred and notes the quality of Elderland Ormsby Mildred Rose.

Pink's farm is in hill country with large forests mixed

with productive fields. The struggle, he said, is to practice soil conservation to keep the soil on the hills.

His plan: keep as much land as possible in grass and "just don't work the soil to death." Pink used the erosion protection method of no-till planting on 40 acres of corn this year.

Conservation methods have worked and he and his brother have 282 acres of corn and are able to market 22,000 bushels a year, primarily to neighboring buyers.

Some of the corn in the bottoms this year is in the seventh leaf or 8 to 10 inches high. Planting started April 23. Last year timely rains gave Pink a yield of 120 bushels per acre. In the 1988 drought, yields averaged 59 bushels per acre.

"Vistors to our farm will find this excellent alfalfa country. That means there is plenty of feed and there should be more farms with beef and dairy cattle and sheep," he said.

But the problem for farmers here is being able to hire the help they need. New plant and industrial job openings are tough competition. Farmers find it more and more difficult to stay in business.

The farm problem is that returns are not enough to pay competitive wages. "It is going to be hell for farmers to get seasonal help," he added.

As for government and its farm relations, Pink faults Washington for trying the keep its "cheap food policy and not look down the road as to what could happen."

Pink worries that the government is killing the incentive farmers need to stay in business. To Pink what is happening in the U.S. is just another way of doing what Stalin did in Russia and that was to kill off the farmers.

Pink knows of what he speaks. He is chairman of the Farm Credit Board for Southwest Wisconsin.

Stoughton Couple Honored for Farm Youth Leadership

November 22, 1992

John and Darlene Arneson, a Stoughton farm couple, share special Farm and City Week honors for their "inspirational and dedicated leadership on behalf of farm youth and their role in building a productive basis for family farms."

It is a role, according to Arthur L. Kurtz, who presented the Agricultural Service Award from Madison's Downtown Kiwanis club, "which not only extends to their community, but throughout the state and nation."

Both John and Darlene are UW-Madison graduates in Agricultural Education from the College of Agricultural and Life Sciences. And both were high school agricultural teachers — John, at Rio and Darlene at Cashton and Cambridge. While in high school, John was state Future Farmers of America president, and Darlene, from Verona High School, was the state sentinel.

"We both tried off-farm jobs and just were not happy until we moved back on the farm and returned to production agriculture," John said.

They were part of the trend, John said, developing around the nation as more and more young college graduates are returning to the farm to pursue their interest in agriculture.

He said the number of farmers in the nation continues to decline with less than 2 percent of the U.S. population directly involved in producing food and fiber.

John, 35, said as a result of this change in farm numbers, "we need quality, well-educated people because of the technological changes in agricultural production."

John and Darlene Arneson

The Arneson farm is a partnership arrangement, with John as herdsman working with 60 Holstein cows.

Darlene is vice-president of the Dane County Farm Bureau and a member of the state policy development committee of the Wisconsin Farm Bureau Federation. In her committee role she will present the resolutions to Farm Bureau delegates and explain the policy statements to delegates wanting to understand the impact of farm organization recommendations.

Both Arnesons are delegates in Wisconsin Dairies

Cooperative.

The couple was married in 1987 and they are raising two sons, whom Darlene describes as "bundles of energy" — Lars John, 3, and Jens Olaf, 1.

Darlene uses the farm as an outdoor laboratory, hosting second and fourth grade school children in the Farm Bureau's Agriculture in the Classroom program.

She uses every forum she can to help non-farm audiences understand agriculture because she said she believes full understanding of farming "will help us face our future on the farm."

In a tribute to the Arnesons, Professor Joe Von Elbe, UW food scientist, said Americans have an abundant food supply which is economical, nutritious and safe for a very simple reason: "We have the most productive and efficient agriculture in the world. Although we give science and technology much of the credit, the biggest resource in agriculture is the people, like the Arnesons, whom we recognize today."

DeForest Anhalt Family Hosting Dairy Breakfast

June 14, 1992

DEFOREST — Starting about 5 a.m. on Thursday, 5,000 people or more are going to have a day to remember. The crowd is expected to gather June 18 at the Anhalt Farm, on Windsor Road, a short distance east of Highway 51. Tickets are $3 at the farm.

This is the spot for Dane County's annual June Dairy Month Breakfast on the Farm.

Do you know what happened to bring this crowd to eat and to visit the Anhalt's Farm — and also have some fun? It was a visitor from India, a guest of Wisconsin Dairies,

Baraboo, who was visiting Wisconsin to learn how state farmers used the cooperative system in their farm living and farm operation.

Art, Helen, and Randy Anhalt did such a superb job telling the story of their fourth generation family farm that Howard Mack, Wisconsin Dairies' chief fieldman, decided the Anhalts should be nominated as the 1992 host family.

Mack knew that the Anhalts' farm would be the place for breakfast, but the Anhalts didn't learn of their responsibilities until Mack announced it at Wisconsin Dairies district meeting.

The Anhalts have had a partnership farm operation since Randy, 29, joined his parents on the 140-acre dairy and hog farm, which also has 140 rented acres.

There are some exceptional practices. For instance, the entire farm has been laid out with contour strips. Water and soil runoff from the fields, Art explains, "amounts to practically nothing."

They raise 350 40-pound feeder pigs to market weights of 240 pounds in holding areas where each one is covered with a single sprinker.

One reason breakfast visitors can expect to have fun is that Art and Helen know how to have fun.

For instance, they are members of the Diamond Squares square dance group and they'll have a dance floor for club members who will show visitors their skills with a few dos-a-dos.

So visitors can get their breakfasts served quickly, the Dane County Dairy Promotion group, breakfast sponsors, will bring in 80 farmers and their wives to serve the food and all the trimmings — like a fresh strawberry sundae, for instance, to top off the meal.

Randy, 29, not only is a skilled dairy herdsman and operator of field equipment, but he also has acquired skill as a landscape designer. Around the exterior of the Anhalt home there are samples of his landscaping efforts.

Randy Anhalt with his father Art and mother Helen

Their hogs are known, Art says, as the mortgage lifters on the farm, but the pride and joy is the dairy herd of 40 cows.

That may sound like as small herd in a day of herds in the hundreds, but this Holstein herd is exactly what they want.

Half of the herd is the traditional black and white Holstein and the other half is red and white. The average production for cows in the herd is about 18,000 pounds.

Furthermore, every cow is named and all have their favorites. Randy depends on the two near the front of the barn — Jackie and Fern.

As generations of cows go through the barn, is it a problem for the Anhalts to find a name?

"No," Randy replied. "I think I have every cow name there is from a listing of all the known cow names published in a 1966 copy of Hoard's Dairyman."

And visitors quickly note the large number cows having names starting with the letter, "D," to designate their reliable cow family. It is a closed herd and Art bought only a few heifers during a period when all the calves born were bulls.

Art is an independent dairy farmer who does not rely on participation in government farm programs to sustain his operation. He said he is proud to be a member of Wisconsin Dairies, which hasn't sold a pound of milk in the form of a dairy product to the government for three years.

This belief in one's own ability to operate the farm as they wish stems from German heritage going back to 1899 when the family came to America from the Erfurth area of Germany, where they had farmed.

His father bought the farm in 1914 and Art took over the operation in the 1940s. Art and Helen have five children: Randy, Linda, Nancy, Karen and Debbie.

Vigo Farms Star of Progress Days

July 7, 1991

BROWNSVILLE — There are 32,162 dairy farms in Wisconsin, but one in particular will be special in America's Dairyland this week.

It is Gordon E. Berg's adjoining Vigo Farms — Vigo I, Vigo II, and Vigo III, just off Highway 49, a short distance west of Brownsville in Dodge County.

Berg is host farmer for the 1991 Wisconsin Farm Progress Days opening Tuesday and continuing through Thursday.

Why should he undertake a massive program of farm improvement — building three new barns, a new home for his farm manager, Douglas Bloch, his wife, Linda, and family, and a new calf raising facility he designed for his most valuable crop?

This Farm Progress event is a way for a patient and

talented person with exceptional cattle breeding skills and dairy systems management to say "thank you" to the people in American agriculture.

"Dairy agriculture has been good to me," Berg said, "and I wanted some way to show my gratitude. This Farm Progress event is my way of giving something back to agriculture and the people."

For those rural and urban residents who never have had the chance to see excellence in dairying — Holstein breeding and farm facility design and art — Berg is opening his doors for a memorable, inspirational visit.

Farm Progress Days is a free show for 150,000 persons and there will be plenty to see in a 60-acre tent city with 800 exhibitors, plus field demonstrations of forage harvesting — from 8:30 a.m. to 5 p.m. daily.

But for those visitors who want to see what daily devotion and success can mean to one farmer — and to the reputation of the state and the Holstein breed as well — they never will forget this tour of Berg's Vigo (pronounced V-eye-Go) Farms' successful operation.

The named Vigo comes from "Vi" for Berg's sister Vivian and "Go" for Gordon.

Just remember, as you enter those barns, Berg can show you 14 cows side by side classified as "excellent," the top score in the Holstein breed. His consistency also is shown in the 32 cows classified as "very good."

His ranking as a breeder is even more sensational because he has placed 110 bulls in artificial insemenation service — a national record.

His favorite bulls, both among the 18 bulls at American Breeders Service, DeForest, were Vigo Raverion and Vigo Citerion.

"They really helped me and meant a nice income for me," Berg said modestly. And remember, he still had 108 other bulls working for him in stud across the nation.

His two milking herds have a rolling average produc-

Gordon Berg

tion of 22,279 pounds of milk and 807 pounds of butterfat. Who is this man of means and manor? He was born on Vigo Farm I and still sleeps in the bedroom of the meticuously kept farm home where he was born, the son of Elmer and Ardis Berg.

It was a good farm with normal prospects until Elmer bought registered Holsteins as 4-H dairy projects for Gordon and Vivian. Those registered animals were the signal of success to come at Berg's farm. Vivian went to school in Illinois, married there and reared her family in her new home state.

Gordon never married, but stayed on the home farm to work with his father. But he never worked for wages. Instead his compensation came when his parents transferred one-third ownership of each Holstein to his name.

His generous reward came as the offspring of the herd were sold.

Because his income came from the sale of cattle, Berg made up his mind early that he would develop the best cattle to gain top income. This concept never changed and Vigo Farms and Berg prospered.

He still doesn't separate himself from the work on a dairy farm and Berg is up at 4:30 a.m. every day to help his hired man of 40 years milk the black and white herd. His farm manager milks the red and white herd.

Berg recalled that he has attended nearly every Farm Progress show in Wisconsin — but all had the barn doors closed so the dairy herd could not be seen.

His doors will be open.

Berg is a barn designer, artist/ decorator in barn coordinated color, innovator of such ideas as an elevator to take the farmer to the second floor of the barn so as not to take up valuable room with a stairway in the feeding alley.

For color, note the barn at Vigo III, which is in a red and white pattern for his red and white herd, and Vigo I is black and white for the traditional Holsteins there.

Also, he has an 1,800-foot white plank fence running between Vigo III and Vigo I. Half the fence posts are painted red, the other half are black to hold the white planks.

Why did he develop the red and white Holsteins? Berg said he receives 25 percent more on the price of a red and white bull because they are rare and hard to find.

Visitors should seek out and meet Berg — the man of means and manor.

Speaker: Customers Going Online

June 16, 1996

MINNEAPOLIS — Retail food stores are in for a shocking change.

At the recent seminar and exposition of the Madison-based International Dairy-Deli-Bakery Association (IDDA), Carol Christison, executive director of the association, said in her annual report that half of the stores today "will not exist in 10 years."

And she really made a telling prediction to the thousands who listened at the Minneapolis Convention Center by asking the men and women to take a look at the people sitting in front and to the side of them.

"The odds are that if they're both retailers, one of them won't be in business in 10 years. And if 50 percent of the retailers won't be in business in 10 years, that puts suppliers at risk too," she warned.

This is what Christison called the battle for the share of the stomach and said it could be called "Store Wars." One of the reasons for the fallout of supermarkets, she continued, is technology.

And all 5,000 registered food business members were there "to make sure they will be included in the 50 percent club" — which is the half that will be sharp enough to survive.

What is happening in the world of technology, which Christison explained, is a combination of pen-based computers, cellular phones, fax machines, pagers, scanners and e-mail.

What that collection is doing, Christison explained, is

making "you available to your customers at any hour of the day or night."

All this is based on the telephone, a 100-year-old device, which suddenly has been updated and expanded through the Internet.

She told of a recent study by the Smart Shop, which showed that changes in retailing are coming "not from the retailer, but from the consumer."

There is only one thing, Christison said, that consumers hate worse than shopping — and that's standing in line. On the Internet, there's no line.

Consumers have a lot of wants — like convenience, ease of shopping, information and low prices. The food store's role?

Christison said it means creating innovative products, making it easy for consumers to replenish their food supplies through the phone, fax or online ordering and home delivery.

Moreover, stores offering these services will start to include such things as banking services.

Some stores with home shopping provide home delivery themselves; others connect with a service. There are three shop-at-home providers in the nation, including one, called Peapod, in Evanston, just north of Chicago.

A comment on food ordering from Bill Gates of Microsoft, is that by the year 2005, supermarkets will lose about one third of their business to electronic media. Other experts predict the home delivery market will top out at around $14.5 billion annually.

"Either way, the smart money says that if your customers aren't in line they'll be online," Christison said. And what a change, according to Gates (the richest man in the world), who envisions the information Superhighway as leading, not to a global village, but to a virtual marketplace.

Brown Swiss Inspire Memories

October 6, 1991

When Ken Salmon goes to World Dairy Expo, the international show closing today at the Dane County Exposition Center, he heads for the Brown Swiss barn and the flood of memories of his and his pioneer family's days in dairying.

At the Voegeli Farm's Brown Swiss show string from Monticello, he quickly recalls the bull calf, a son of the Design bull purchased by his father 50 years ago for not much more than $50.

It was one of the family's best decisions, because the Son of Design from the Voegeli herd produced progeny that returned up to $800 or more for each animal sold.

"I remember when we dispersed the herd with an auction in 1958, one of the first cows through the sale ring got an $800 bid and my neighbor told me he never thought any animal was worth $800.

"And he had Holsteins," Salmon chuckled.

Howard Voegeli recalled that for big shows, like the International Dairy Show in Chicago, he would ask Salmon to join the Voegeli show string because he was a born Brown Swiss exhibitor.

Most of Salmon's memories were of his grandfather, Sam Salmon, who acquired the 193-acre home farm on Highway 39, five miles east of Mineral Point, in the 1800s and ran it until Ken's father took over in 1912.

"My grandfather," he recalled, "was known as the neighborhood veterinarian" because of the way he could concoct remedies for ailing animals. And who knows, he may have helped some ailing people, too."

His grandfather put together one substance called

"Black Oil," and on the little label was the notation: "good for man or beast."

"For years we had demand for the medicine and we used to mix it in an earthen crock and had a paddle to stir it with. The ingredients included turpentine and witch oil and you name it, and a gum-something," he recalled.

Salmon still has his grandfather's recipe book of remedies.

The senior Salmon was quite a horseman. He can recall how his father told of the times his father put his fingers in his mouth, whistled and 20 colts would come down off the hills.

There was a small race track on a piece of level ground on the farm and his grandfather kept contact with his customers because he was "a pretty good smithy and a metal expert on the forge."

Salmon counts his good fortune in life because he had a very progressive father, who would attend University of Wisconsin field days where UW professors would tell him things to try on the farm.

"And the professors would add, 'give it a try and if it doesn't work, forget it,' " Salmon added.

Some ideas did work and the Salmon farm was one of the first lime grinders even though his neighbors asked how stupid he could be because there was an abundance of limestone in the area.

But his father, Will, went along with the UW idea and added the ground lime to his soil. This balanced soil enabled him to become the first farmer in the neighborhood to shift from clover and timothy hay crops to leafy, protein-packed alfalfa.

"That's because my father listened to the professors and found he could improve his forage crop. He also put in some of the area's first terraces and strip crops for soil conservation," Salmon continued.

The Salmon farm was a neighborhood stopping place

because of all the experimental work on crops and machinery.

"My mother never knew how many men she would have for dinner," Salmon laughed, "because people would come in for a look at things new. Even the mailman, who delivered his mail with ponies, usually stopped for dinner."

After dairying, Salmon was a membership worker for Pure Milk Products Cooperative and, after its merger, did the same work with Associated Milk Producers.

His territory covered the ridge and valley region of northern Illinois and he wonders today how he ever found some farms in the rugged hill country.

"But those were friendly people living on rough farms and they persisted and survived and came out pretty wealthy," Salmon said.

Reluctant Farmer's Wife Now Leads

November 3, 1991

DEFOREST — Once a year Janice Walton, of Simmental Valley Farm, fills a bus with shoppers and, naturally, heads for Madison stores.

Her busload of shoppers include men and women from 10 or more countries attracted to the annual World Dairy Expo. This year Expo had 2,600 persons registered from 60 countries.

Walton, wife of Bob Walton, chairman of American Breeders Service, DeForest, is chairwoman of World Dairy Expo's International Cultural Tour, which is part of Ilene Farr's Craft Forum Committee at Expo.

It is a prestige tour that gives the international visitors a chance to become acquainted with farms and farm homes, the special places around Madison, UW-Madison, the State

Capitol and the retail businesses of Madison.

The tour's frosting on the cake is a luncheon for Wisconsin's guests at the governor's residence on Lake Mendota in Maple Bluff.

And Walton said she loves this volunteer service. Not only is it fun, she said, "but the choice position of all the committees serving World Dairy Expo is the one I have."

Janice Walton

Earlier, the tour was called the Women's tour, she explained, "but now we are finding more and more men want to go. We are encouraging them; we felt it was too bad that they came all this distance from their homes and didn't see our beautiful Madison area, its people and its institutions."

Walton, who has done extensive world travel, said she often thinks of how much she would enjoy such trips if she had a chance to visit farm homes.

When she started the tours about 10 years ago, the guests visited farms but not a farm home. "The visitors let us know they would really like to visit a farm home," she said.

Even Walton's home was included on the tour when the group passed her farm en route to a neighboring home and she mentioned that they would be passing her home on Viburn Road.

That quickly changed the schedule, and the group wanted to see the home of their host.

"All right, if you promise not to look too carefully in the corners. I was hardly prepared for this visit," she said.

As for the shopping portion, the group this year went to West Towne Mall. Some of the party made lists of the things they wanted to take home. The merchants helped plan the excursion and helped them find the presents.

One visitor went home with a skateboard under his arm. The end of the shopping trip, Walton said, is a show-and-tell feature in the bus, in which each visitor gets "ohs" and "ahs" for their purchases.

Walton served with co-chairwoman Donna Nesbitt, Fort Atkinson, for the early years of the tour, and now with co-chairwoman Aleide Bos, also of Fort Atkinson.

Her interest in acting as a host traces to her days at Iowa State University, where she took her graduate work and was an assistant professor in child development.

It just happened that she was dating an Australian, who was a friend of Bob Walton. Well, the Aussie became an international inspiration, and Walton, who was getting his doctorate in genetics and statistics, won her inspiration.

They were married in 1959 and moved to Madison in 1964 as the ABS headquarters was being constructed at DeForest.

When Bob Walton asked her to marry him, the city woman from a banker's family in Sioux City, Iowa, exclaimed, "but I don't know a thing about farming."

Bob Walton's reply is on a wooden plaque in their farm home kitchen. It says, "But Jan, I'm going to be a professor, not a farmer."

Today Walton is midwife to the beef herd of Simmental cows, which have their calves at all hours. But she is ready. How big is their herd? Actually the cows and young stock total 68, but Bob Walton says, "no more than my wife can handle."

They have three children, Cindy, wife of dairy farmer Stephen Pederson, south of Madison; Bob Junior, training in Iowa to be a chiropractor; and John, finishing a master's degree in business administration at Edgewood College.

Volunteers Spark Craft Forum

September 22, 1991

When World Dairy Expo opens its 25th year on Wednesday at the Dane County Exposition Center, Ilene Farr and her group of 20 or more volunteers will offer a new chapter in Expo's Craft Forum.

Farr is Expo's chairwoman of the Craft Forum Committee, responsible for the Craft Forum, one of three scheduled daily features of the five-day Expo program continuing through Sunday.

Throughout the year, committee members plan and work to attract the talent visitors want to hear and see at the Expo. Farr, who has been in charge of the Craft Forum since 1979, said it really is "a great group of volunteers that is responsible for the way the Craft Forum has been sustained since Expo started in 1967.

The Craft Forum program format went through a major content change a year ago, Farr said, and the way its daily audience accepted it the first year indicates they have the kind of program women enjoy.

The favorite sessions for women are those with humorous speakers, like Floyd Doering on Wednesday, State Journal columnist George Hesselberg on Thursday, and Orion Samuelson, WGN farm director, on Friday, select cooking demonstrations and style revues and clothes and home features daily.

The program is free and in the same area is the Purple Cow, a dairy oriented gift shop, which also is the responsibility of Farr's committee.

Farr said the name, Craft Forum, describes a lot about the program. In addition to the daily stage programs, 46 crafters will be at the shows Wednesday through Saturday,

Ilene Farr

from 9:30 a.m. to 4:30 p.m. in the Forum Building.

On Sunday, the large auditorium space in the Forum Building will house 85 or more crafters from Wisconsin and elsewhere in the Midwest. The program becomes Craft Day and lasts until Expo closing at 5 p.m.

"There is one audience we haven't consistently attracted and that's an urban audience from Madison and its suburban area. But we believe we have a daily program that would please people living there," she said.

Farr is an Iowa native and was raised on a beef, hog and corn farm near Marcus in northwest Iowa. As a program official of World Dairy Expo she had to have some exposure to dairying and she got that by milking the family cow late each afternoon.

Over the years were her cows registered Holsteins?

"No. But remember, this was an Iowa livestock farm and they were just cows." she laughed.

After graduating from high school, Farr took nurse's training at St. Joseph's College in Sioux City, specializing in orthopedics. As a registered nurse, her first position was assistant head nurse in the orthopedic section of St. Joseph's hospital.

It was on a blind date in 1949 that she met Charles Farr, who was to become her husband a year later. At the time Charles was a member of the milk market administrator's staff.

Farr took a position with the Veterans Hospital in Lincoln, Neb., when Charles completed his master's degree studies at the University of Nebraska. Charles took a job in the Twin Cities as a dairy economist, and the Farrs moved there and reared their family. They have lived on Madison's West Side since 1962.

The Farrs have three children and have annual trips to visit them and their four grandchildren in such far off places as Alaska, where their son is a major in the Army. Another son is in California. Their daughter lives in Matteson, Ill.

While working with volunteers at Expo, Farr also is a volunteer worker with Good Shepherd Lutheran Church and at Oakwood Lutheran Home.

Non-Native Knows a Whole Lotta Folks

January 27, 1991

Myrna Sue Jones has been living in Wisconsin for only 10 years, but she probably knows more people in Dane County and the state than most natives.

What's more, she remembers their names.

Jones, who lives with her husband, R. Dale Jones, on Sunnymede Farm, Oak Park Road, Marshall, got acquainted with the people of agriculture, agribusiness, the professions and those in commerce as a volunteer.

She was born in northeast Iowa, daughter of a Fayette County dairy farmer.

Myra Sue Jones

Jones had her share of farm experiences, mostly with hogs or chickens and learned a lot, both as a farmer's daughter or a farmer's wife. She is the mother of five children, three of her sons are farming with their father in Iowa, one in the Madison area and a daughter in Michigan.

Years ago, when she was a visitor at World Dairy Expo, she met R. Dale Jones, an Indiana dairy cattle breeder. They were married 13 years ago.

Her volunteering started in Iowa 4-H in various leadership positions. Her enthusiasm for youth in 4-H brought her a job as a paraprofessional of Iowa State University as a 4-H coordinator in her home county.

She became known for always triggering something new. When two members of a 4-H club in Iowa said they wished they could show their goats at the county fair, she negotiated.

"You help me find another family or two with goats and if we get 10 head, we'll have a 4-H goat show," she said. They found two other families and Jones organized the dairy goat show. It is still going today.

She likes the concept of 4-H because it is important for young people to be in a program in which parents and children can work together.

In the Marshall area she worked with the Medina Marvels 4-H club, where young people work closely with their parents to get the feeling of work and responsibility.

"I have had instances in the 4-H family in other states where parents and children never worked closely together until the children were part of a 4-H project," she said.

When she visited World Dairy Expo regularly she became one of the scores of vounteers and wound up decorating the Memorial Coliseum for cattle shows.

She organized a program for fourth grade children to take guided tours of Expo to learn more about dairying and agriculture.

Jones now is working for Expo as public relations coordinator. "In this job you learn to take constructive criticism and it helps," she said.

Her volunteering moved her to the Dane County Dairy Promotion Committee, of which she became president. That led her to participation in the annual Cows on the Concourse program where she wants children to realize that milk comes from clean cows kept in clean barns, and living in a clean environment.

The list of things to do as an advocate for youth and agriculture and its agribusiness sector found new expression in the Dane County Agribusiness Delegation, which she helped organize in 1984 and now is serving as president.

She also is district president of AMPI Women in the Morning Glory Farms region of Associated Milk Producers Inc.

She has a post with the National Holstein Association's scholarship program and has been a junior Holstein advisor in Iowa and Indiana.

At St. Paul's Lutheran Church in eastern Dane County she is a member and vice president of the church council. Church work is important because Jones is waging a two year battle with breast cancer and has just completed 24 radiation treatments.

She has no qualms about the treatments because "it is

saving my life," she said.

'Also the church council is praying for me. That's a real help — every day— and I need all the prayers I can get," she continued.

"I get cards, notes and even flowers from people I don't know," Jones said.

She also volunteers her time for others who have cancer, helping them know what to expect so they can have a positive attitude to see their way through.

"The most traumatic thing I ever had was losing my hair. I had it down to here," she said, placing her hand at the small of her back.

"But things change. We change. Life changes . . ."

Dairy Expo in Madison is International Meeting Place

December 17, 1995

There isn't a farm show in the nation that brings in the number of international visitors that World Dairy Expo brings to Madison each year.

Most of the dairy visitors from abroad have made Madison the world center of dairying and they come to stay current with the changing technologies of dairying, in equipment and the trends in dairy cattle breeding.

They also meet U.S. and foreign business representatives who will be their sources of supplies, ideas, new equipment and cow care, which can improve their dairy business operations.

This year, final figures show that the five-day Expo brought in 3,706 guests from 78 countries to give Expo both a record in the number of countries represented and the number of visitors from those countries. A year ago, 71 countries were represented by 3,165 visitors.

As usual, Canada sends the greatest number of visitors — 741; Germany was second with 276; Japan, third, 227; Brazil, fourth, 191; Argentina, fifth, 159; Switzerland, sixth, 157; The Netherlands, seventh, 150; and United Kingdom, eighth, 145.

The international visitor system is run by Kathy Stoikes, who has conducted the program for 10 years. Stoikes is a program assistant in the International Agribusiness Center, a Markets Division unit of the Wisconsin Department of Agriculture, Trade and Consumer Protection. She has a supporting registration staff of agriculture department specialists.

As an organizational expert, Stoikes knows just what is going to happen and the kind of questions the international group will want answered.

A highlight for visitors includes visits to Wisconsin dairy farms and dairy equipment firms.

Stoikes said that because World Dairy Expo fills up Madison hotels and motels, the international visitors experience other Wisconsin communities since many of them stay up to 50 miles from Madison.

The demand for interpreters to help the foreign visitors is interesting, Stoikes said.

Therefore, she needs skilled interpreters from UW-Madison and volunteers who can converse with visitors in the 10 main languages, including French, Spanish, Portugese, Italian, Chinese, Japanese, Polish, Russian, Ukrainian and Dutch.

Stoikes also plans major international events such as the popular reception for foreign visitors, where they get a chance to sample a variety of Wisconsin foods and beverages.

Art Nesbitt, Fort Atkinson business executive, who tapes interviews with international guests, asked a cattle breeding authority from the Netherlands why he comes to Expo each year.

"I am here every year to do business, and meet my customers from Germany, a major market for my business," the cattle breeding authority replied. "But your firm is only a few miles from the German border. Why Madison?" Nesbitt asked.

"Because Expo is where my customers are," the visitor replied.

There are scores of dairy businesses from overseas at Expo each year, many of them to show and sell their products exclusively to international travelers at Expo. The entire third concourse at the Memorial Coliseum is reserved for international guests and businesses wanting to meet them.

The people from 78 countries are kept informed of what each annual Expo program will include since Kathy distributes program information to a long list of potential international visitors through annual mailings.

Canada Journalist Comes Home for Visit

September 29, 1991

DEFOREST — When the big Canadian dairy delegation of more than 600 men and women arrives in Madison this week for World Dairy Expo, one of their number will be coming home.

Bonnie Cooper, who has lived and worked in Canada for 18 years, will be back in Madison and at Moss Oak Holstein Farm, DeForest, where her parents, Maurice and Geraldine, and brother work together in the dairy operation.

Cooper is editor of the Holstein Journal in Toronto. The journal is the principal dairy breed publication in Canada. The monthly magazine has a large international circulation and is published with both English and French sections.

Cooper travels extensively and visits dairy activities and will get to most Canadian provinces each year.

It is a competitive world because the Canadian Holstein organization recently has started publishing its own magazine and the Journal lost its contract as the official Holstein publication.

She is in Canada as a "landed immigrant," which means she is still an American citizen, which means she can pay Canadian taxes and report her income to the U.S. Internal Revenue Service, but she can't vote in Canada.

Bonnie Cooper

But, as the length of her Canadian stay indicates, she does like Canada, its people and her work because it enables her to combine her two farvorite worlds, "my journalism and my love of Holsteins."

She acquired the love for Holstein cattle on the home farm, which has a registered Holstein herd, and through the competition at the Dane County Junior Fair, as a dairy project member of the Hollow Creek Comets 4-H club. She was 10 when she raised her first Holstein calf.

Her love of journalism started with her degree in agricultural journalism with emphasis in Dairy Science course work. She also worked with the Cheese Reporter, a Madison-based publication.

Cooper was at the UW-Madison during a time, 1969-1973, when women were slowly being admitted as majors in different courses, but she did not get her wish to have a double major, including one in dairy science.

Only two women before her had completed such a dairy science major, but when Cooper asked for permission to do the same, the department recommended that she transfer, instead, to home economics.

That's not what she wanted. Cooper was disappointed because, as a senior in DeForest High School she visited with both women who were enrolled as dairy science majors and was encourged to do the same.

But Cooper was a determined and didn't let that decision bother her. As an alternative she stayed in agricultural journalism and then scheduled all the dairy science courses possible.

Was she good in dairy science? She received an award for her dairy education and that seemed a good indicator of her ability.

Also, early in her professional career she showed that a woman could compete. Just two years after getting the Canada magazine writing position, she was named editor of the Holstein Journal.

Cooper was the first woman employed at the magazine's senior staff level.

Moving to Canada is not like moving to another state, she explained, it is a different life, although there are some similarities to the United States.

She thought the job would be a good place to get started and she probably would come back to the states. Somehow that job went on for 18 years.

She looks back at her high school education and her thoughts of one day running a farm, but that wasn't a time when women went into farming, "it just wasn't the thing to do," she said.

That also led to her decision to do something that would keep her close to Holsteins.

Joyce Bupp

Clarence Boyke

Doug Wilson

Dairy Expo Awards Cross Gender, National Lines

October 3, 1993

What do you say when you receive a top honor of the dairy world?

Four honorees had the chance to say their piece last week at World Dairy Expo's awards program.

Recipients were Joyce A. Bupp of York, Pa., Dairywoman of the Year; Clarence Boyke of Fond du Lac, Dairyman of the Year; Doug Wilson of Shawano, Industry Person of the Year; and Verghese Kurien of India, International Person of the Year.

Bupp and her husband, Roy, have 450 registered Holsteins, she milks seven days a week, helps with field work and handles the farm bookkeeping. She also has time in her 6:30 a.m. to 11 p.m. daily schedule to write columns and stories for newpapers and dairy publications.

Her message was dedicated to dairy farm women, who, she said, "for years and years have been helping make this industry one of the nation's finest parts of agriculture."

She praised Expo for recognizing the role women have on farms and said she accepted the honor, "not for me, but for my family and for all you women, who have ever milked a cow or ever got up in the middle of the night and had to get the heifers off the road and back into the pasture."

Boyke, who with his wife, Virginia, developed a Holstein herd with state and national ranking, praised youth who have the pep, vigor and get-up-and-go needed in dairying today.

His message was on developing a positive promotion and marketing program since "I do not believe you can produce a product and not sell it." He wants to see merchandising and advertising programs strong enough to keep dairying "the biggest industry in the world."

Wilson, chief executive officer of 21st Century Genetics in Shawano, emphasized World Dairy Expo's benefits to dairy farmers and Madison.

He travels extensively around the world for his cooperative and said, "I have never been in a country where they did not make a reference to the first week in October as the week of World Dairy Expo in Madison, Wis."

He said people had to work to "keep talented and bright young minds enthused about the dairy industry. You've made Madison America's Dairyland forever."

Kurien is chairman of India's National Dairy Development Board and developed a cooperative system in India that produces, processes and markets milk. It involves 5 million dairy producers and 50,000 dairy cooperatives providing a stable and profitable market for dairy farmers and milk for consumers in 450 cities and towns.

"We have a system founded on research, production and marketing that makes it the largest agricultural development in the world," he said.

He was a student at Michigan State University and came to the Fort Atkinson and Lake Mills area to work in a modern packaging factory.

"They say every good man has to have a good woman behind him and that is also true in India," he said in a salute to his wife who attended the honors program.

Chapter 7

Agriculture Departments: Wisconsin and USDA

Monticello Grad Began her Rise with Sheep, FFA

December 8, 1991

It's time to meet Wisconsin's Mary Keough. Today she is an agricultural statistician working in the National Agricultural Statistics Service on the fifth floor of the South Building at the U.S. Department of Agriculture. Each month she announces the basic price of milk.

A month from today, however, she will be manager of dairy economics at Kraft Foods in Northbrook, Ill. And it wasn't many years ago that she was a farm girl starting a Hampshire sheep flock with a ewe and lamb from Werner Zimmerman of Wyota, on a dairy farm in Monticello.

Keough is still in her 20s and has such admiration and respect from her peers in the dairy industry there are those who proclaim she probably will be the first woman to become U.S. secretary of agriculture.

"Oh, please give me a few years," she pleaded, while showing her pleasing Irish smile.

But you can get a hint of her credibility among dairy industry leaders, who, upon learning of her decision to join Kraft, saw her and in numbers told her, "We didn't know you were looking. Why didn't you let us know?" "I didn't go looking," she calmly replied.

But where did this young woman develop her professional foresight?

Keough was born in Darlington and lived on rented farms there until her parents purchased 160 acres in the Monticello area.

She developed a superb record as a member of the

Future Farmers of America chapter at Monticello High School from which she graduated in 1981. She was elected a state FFA officer.

The FFA became her stepping stone — both domestic and international — in the world where people look for youth with energy and educational desire.

Of course she was encourged by FFA leader Arnold Cordes of Madison, whom she admires.

Keough's accomplishments may leave you breathless.

Mary Keough

She won the state FFA (prepared) speaking contest and went to compete nationally at Kansas City. Among those who heard her was the dean of agriculture of Texas A&M.

The dean was the first to learn that with Keough you don't wait to see if she was looking, he immediately offered her an A&M scholarship.

Keough wasn't ready for Texas, so after meeting Agricultural Economist Robert Cropp, she chose UW-Platteville to begin her studies.

During her 1 1/2 years at Platteville, Cropp awarded her an internship at the Chicago Regional Milk Marketing Order. That experience gave her an important dairy perspective through payroll accounting.

Keough then used her FFA ties to travel and learn, first going to Germany, where she lived on a dairy farm in 1982 and 1983 and used the experience to get a firsthand education about the start of a European Community milk supply management program.

Then in 1984, she was off to Japan, where she also lived on a farm producing massaged Kobe beef. Her host family also allowed her to work on the neighbors' dairy farm.

Now she can hold her own with the German language and can dazzle you with many Japanese phrases.

"When I came back from Japan I was 21 years old and focused on where I wanted to be," she said.

So she picked up her Texas A&M scholarship and, although she majored in agricultural journalism, she started building a base in economics with Ron Knutson, who, like the dean, saw her potential and invited her to work in the dairy policy field.

Knutson has the ability to get grant money, which in turn was used to finance studies for Keough on California dairy farms and at milk marketing order hearings.

Then she needed more depth in domestic issues. In the summer of 1986 she was selected as an intern by U.S. Secretary of Agriculture Richard Lyng. He asked her where she wanted to work and she chose the Foreign Agriculture Service.

After getting her bachelor's degree at Texas, she applied for and obtained a Fulbright scholarship to go to Germany for a year to study milk supply management plans, which were being discussed in the U.S. As a result of her German dairy farm experience with the management program and her academic studies in Germany, she has developed expertise in dairy policy development.

When she returned, the Foreign Ag Service asked her to join them to study the free trade agreement proposed between the U.S. and Canada.

She also took time out and used another FFA opportunity to learn about and communicate with 225 young people from around the world who cruised the seas on a Japanese "Ship for World Youth."

By that time her mentor, Knutson, secured more grants for the Agriculture and Food Policy Center and he invited Keough back to study for her master's degree in agricultural economics. So now she is both an ag journalist and an ag economist.

She completed her Texas studies in the fall of 1990 and was quickly off to New Zealand for a visit "before working for the rest of my life."

'Alice' Builds Understanding Between Farmers, Consumers

March 22, 1992

Last week was Agriculture Week in Wisconsin and throughout the nation, which provides a chance for consumers and farmers to know each other a little better.

But for eight months, Renea Troeller, from a dairy farm near Rubicon in Dodge County, has been making an effort to help the two groups better understand each other.

Troeller, 23, is Wisconsin's Alice in Dairyland. She is the 44th woman to carry the title in a program that has been continuous since Wisconsin's Centennial Year in 1948.

The uninterrupted selection of Alices shows how successful the program has been.

Troeller is a one-year employee of the Wisconsin Department of Agriculture Trade and Consumer Protection. She is paid $20,000 for the year and has the use of a state car.

She has been bringing the message of consumer-farmer understanding to Wisconsin audiences, including adults and school groups.

Her schedule is filled for six days every week and, on occasion, a seventh day.

And she says she loves it.

As a speaker, she can hold her own, even when sharing the rostrum with the pros as she did Wednesday morning with Gov. Tommy Thompson and State Secretary of Agriculture Alan Tracy.

Incidentally, Tracy saluted Troeller as the State Agriculture Department's "most popular employee, who is doing a great job for the state."

What makes Troeller so successful is that she has that touch of farm moxie, has a good sense of humor to keep audiences laughing and adds experiences gained from a variety of interim student jobs, such as the Farm Credit Service and the Wisconsin Milk Marketing Board. She stays well informed by reading newspapers, books and magazines.

Some time in June — her last month as Alice — she will pass the 40,000-mile mark in her state car, "which never let me down once."

Troeller said one of the big things she has gained in the past year is an acquaintance with many of the small towns on the routes she has to travel.

But her Wisconsin travels don't include the extra 20,000 miles she gained in her flight to Japan and back for an international youth summit.

She was in Japan as one of Wisconsin's 10 representatives in a total of 600 young people from around the U.S. This was a Grass Roots Summit with an equal number of Japanese college students. The program was organized to improve relationships between American and Japanese young people.

But you know where Troeller spent her free time — in Japanese supermarkets where she tried to find out the wants of consumers.

She found out, for instance, that the Japanese prefer beef, but it has to be fatty and have heavy marbling for the taste they enjoy. It is their treat once or sometimes twice a week.

Troeller will be a senior in marketing studies at UW-River Falls when she completes her work as Alice.

She has a philosophy that every once in a while people have to stand back and take a look on how things are going. This is especially true in her marketing work because those successful in that profession have to stay informed of what consumers want.

Her goal is to get her degree next year and then find a position in dairy product marketing, preferably in a city where she also can continue her studies to get an advanced degree.

Alice Grows Up to June Dairy Month Chairwoman

May 29, 1990

Dorothy Farrell, Plymouth, will be back in Madison Saturday with a new title. Just eight years ago, Farrell of the Oregon and Belleville areas, was Wisconsin's Alice in Dairyland serving as a staff member of the State Department of Agriculture's marketing division.

Her work as Alice was to help promote the state's dairy and food industries in 1982 and 1983.

She always had a direct connection with the dairy industry — at the home dairy farm where she grew up, as Alice, and now as an executive of a cheese firm.

On June 2, at the annual Cows on the Concourse program at the State Capitol, she will be back in Madison as Wisconsin's 1990 June Dairy Month chairman.

The selection of Farrell as chairman by the Wisconsin Milk Marketing Board singles her out as a leader in the state's $3 billion a year dairy industry.

As products manager for Sargento Cheese Co., Plymouth, she is in charge of product inventories, pricing and packaging.

Her appeance in Madison will be her first official duty as state dairy month chairman and she will preside at the 9 a.m. to 1 p.m. kickoff for the month-long celebration that is part of the nation's largest food promotion.

"I'll get to my share of dairy breakfasts and banquets," Farrell said.

Lee Gentine, now president of Sargento Cheese, hired Farrell even before she completed her year as Alice. She recalls returning from a food promotion assignment in Dallas and starting her new job the following day.

Dorothy Farrell

Last September she was married to Randy Boeldt, a Sheboygan County deputy sheriff, who now is chief of police serving Elkhardt Lake.

With the use of milk quality premiums, she said, dairy farmers are finally being rewarded for the work they are doing to maintain and improve milk quality.

"The important issue is that we do have consumer demand and acceptance of our products. Once they have tried products they are coming back to get consistant high quality products they expect," Farrell said.

"Certainly the dairy products line from farm to processing plant to market is a team effort," she said.

Farrell is a graduate of Belleville High School and holds the state farmer degree for her excellence in the farming program while a member of the Belleville FFA.

She is a UW-Madison graduate in agricultural journalism.

State Ag Statistician Sees New Mexico's Dairy Growth

September 11, 1994

Roger Stewart is best-known in Wisconsin as the agricultural statistician who compiles and sends out timely data on Wisconsin crops and their national impact.

He is starting his 11th year as a federal agricultural statistician with the Wisconsin Agriculture Reporting Service in Madison. His assignment was made by the National Agriculture Statistic Service of the U.S. Department of Agriculture.

Stewart, a graduate in mathematics from Montana State University, is in his 20th year with the department, having served as a crop statistician in Washington, D.C., Colorado and in his native Montana before being assigned to Wisconsin.

With his year-round dedication as an agricultural statistician, Roger said he still travels each summer as "an ordinary tourist" until he sees something special.

But on his August trip to Roswell, N.M., to visit his wife's parents, his thoughts centered on dairying. New Mexico dairying?

On Stewart's visit, his interest focused on the dairy area south and west of Roswell, a retirement community of 50,000 persons.

This area made a dramatic agricultural change as 50 new dairy farms (called dairies there) and a single large cheese plant suddenly sprang to life on the flat plain sitting over a massive aquifer of water.

"Five years ago, you didn't hear a thing about dairying," Roger said. "It was mainly an area of specialty crop production and a few dai ries."

But since then, it has blossomed into a major dairy region bringing in dollars and providing employment.

Wisconsin has a special interest in New Mexico's dairy operations because early this year several tanker loads of milk were shipped from New Mexico to Wisconsin cheese plants to make up for a milk shortage caused by 1993 rains and floods.

Roger Stewart

That was about the same time that word filtered back to Wisconsin that the New Mexico dairy expansion was so controlled that no one could start a dairy operation unless they had a place to ship their milk. It now has three milk plants.

Now, dairy expansion has made it the fastest growing milk production state on a percentage basis — 21 percent from 1992 to 1993 — and it still is not in the top 21 milk production states.

What about the dairies?

These 1,000-cow herds (compared to 50 to 60 cows per herd in Wisconsin) on each farm are Holsteins. Each dairy is designed exactly the same as if one contractor moved in and built the dairies from one set of plans. Some owners live in homes on the farms, others live in nearby towns.

Most of the producers moved into the Roswell area from California and Texas.

Roger, the tourist, told of his visits to dairies where cattle are handled in large feed lots and spend most of the time under high, long roofs for shade. Some days Roger found the temperature never went below 100 degrees, but there also were comfortable days at 85 degrees.

His father-in-law's garden, for instance, needs his attention from February to November.

The cows need no closed buildings because temperatures are not severe. At times they will drop to 20 degrees. Farms also grow alfalfa and harvest five crops a year. That's still not enough to take care of hay demand, so alfalfa is brought in from states like Colorado and Nebraska.

Cows are milked, 40 at a time and are averaging 70 pounds of milk a day.

The area has a cheese plant now, an operation of Lapino Cheese of Colorado. Lapino Cheese provided the big market the new milk producers needed.

But no New Mexico cheese is sold in the Roswell area, it is all shipped out as barrel cheese for processing or block cheese for cutting.

Stewart and his wife, Gayl, met at Montana State University, where they were married. They have a daughter, Heide, a recent biochemistry graduate from UW-Madison.

New Agricultural Statistician Off to Fast Start

October 13, 1996

The new state statistician of the Wisconsin Agricultural Statistics Service is Robert Battaglia, who is getting to know Wisconsin agriculture at a fast pace.

When he arrived in Madison from New Jersey, where he had the same state statistician's post, he spent countless hours with state and federal members of the statistics staff and with a collection of books that tell the state's statistical story.

But he is a fast learner and today, after intensive study,

he probably knows as much about Wisconsin agriculture and its farmers as anyone in the state.

Battaglia (pronounced "Battalia" since the "g" is silent) is happy to be in Wisconsin with its 79,000 farms, compared to just 9,000 in New Jersey. In addition to being larger, Wisconsin's agricultural base is much more diverse than his former state's, he said.

Wisconsin is what Battaglia regards as a "Big Time" agricultural state, with plenty of statistics from dairy, meat production, canning crops, cranberries and cropland agriculture.

His one hope is that "the statistical information our office provides farmers and businesses proves accurate so it can be used with confidence to help people make their major decisions."

Battaglia is a fast learner, with a specialty as a mathematical statistician, who held two posts in that category in special assignments in Washington and in the Michigan statistical service while working on his master's degree at Michigan State University.

He also has international experiences, helping agricultural officials of Ukraine, Bulgaria, and Haiti. A Ukrainian delegation met with him in Madison last month.

Battaglia gained farming experience in New Jersey as a young man. He particularly remembers helping his grandfather make wine and worked with fruit and vegetable production plots, a common practice in New Jersey.

He also laughs about some farm experiences where he worked as a hired hand. The farm owner's sons, who were to be companion workers, quickly sized up his talent and shifted work to Battaglia, instead of doing the jobs themselves.

He took a forestry course at Rutgers University in New Jersey and showed such a strong interest in agriculture that his department chairman encouraged him to stay with it as a graduate student in agricultural economics, where he received a master's degree.

Later he surveyed New Jersey farmer attitudes in a project organized by Roger Wyse, Rutgers agriculture dean, who now is agriculture dean at the UW-Madison.

He also made a Wisconsin connection on a special project with Daryl Lund, former UW-Madison food science professor, now dean of agriculture at Cornell University in New York State.

Battaglia is now making some of the radio reports and the state can learn how he makes his crop and livestock assessments.

His most recent one was on how Wisconsin did in the October crop report. The report is heard on many Wisconsin radio stations and also appears in state newspaper reports.

Battaglia, his wife, Kathy, and three children live in Sun Prairie. Before moving to Madison, Kathy was grants coordinator for Bucks County Community College in Pennsylvania.

"We enjoy our life in Wisconsin and I enjoy my job as state statistician. We have a good staff of 25 state and federal workers who are keeping up with farmers. They respect the confidentiality of the figures which are tabulated for the reports in all lines of Wisconsin agriculture," Battaglia said.

State Seeks to Expand Dairy Trade with Vietnam

November 19, 1995

The Wisconsin Department of Agriculture revealed its interest in developing international markets for state farmers when it opened talks on dairy genetics with Vietnam officials five months before the U.S. and Vietnam agreed to full diplomatic relations.

Wisconsin moved into a favorable trade position on dairy cattle and other livestock when Vietnam's senior livestock official, Dang Tran Tinh, came to Madison for World Dairy Expo.

Christopher J. Gutman, Wisconsin's international manager responsible for developing livestock trade in Asia and Eastern Europe for the state agriculture department was the guide for Dang.

He noted that Wisconsin is the source of more than 50 percent of all dairy genetics — semen, cattle and embryos.

The Vietnam official speaks English well and was dazzled with the excellence of Holstein cattle, particularly heifers.

Dang made it clear, Vietnam wants to increase milk production for its people and must boost Holstein numbers to build its registered herd of Holsteins and other dairy cattle.

Holsteins are being bred to the region's native yellow cattle as a way of getting more milk as quickly as possible.

The local dairy herds are small in the Vietnam countryside where two, three or four head of yellow cattle are kept on farms to produce milk. The milk is worth 40 percent more when it is moved from farms into nearby cities and villages.

But the country wants to expand dairying into five main areas of Vietnam, he said, in which conditions are excellent, even for purebred Holstein herds to form the country's base breeding stock for milk production.

Such areas include some of the highland regions where temperatures are cooler than in the lowlands.

But the government insists that the emphasis should continue on cattle that are heat tolerant so as many lowland areas as possible can get heat tolerant cattle to keep building dairy herd numbers.

Currently, Vietnam is working with the World Bank and the Asian Development Bank to get basic financing for cattle expansion.

"And we are looking for 1,000 cows to be financed by cattle sellers," he said. He indicated the money was really needed to help with the imports of cattle, semen and embryos.

Dang said cattle and breeding agencies in Australia and the Netherlands are willing to finance some cattle.

And he was not hesitant about saying that the country had a market and needed more cattle.

When Dang talks about the cattle numbers needed, he quickly mentions that they have 20,000 head of dairy cattle now and "want to build that to 100,000 head by the year 2000."

FSA Director Has Big Job with new Farm Bill

June 2, 1996

As government programs change and some jobs are eliminated, Douglas J. Caruso is in a key position, which he describes as "critical to the Wisconsin economy and Wisconsin farmers."

Caruso is state executive director of the Farm Service Agency in Wisconsin, a U.S. Department of Agriculture administrative unit that handles a provision of the 1996 Farm Bill — a major infusion of $803 million in payments to Wisconsin farmers over the next seven years — ending in the year 2002.

This is a payment to ease the transition from decades of using comprehensive farm programs run by the government in order to control annual grain production and prices.

But the new Farm Bill cancels all those basic crop support programs and leaves farmers with the freedom to plant what they want, how much they want and where they will market the production.

In Wisconsin, nearly every working farm has an established base acreage for basic crops, including crops like corn, oats, wheat and barley. Each of those crops have base acreages and known yields, the basic requirement making each farmer eligible for the transition payments.

A percentage of the base acreage and yield will determine the size of the payment each eligible farmer will receive.

There are 39,000 farmers with base acres on the farms they own and there are more than 54,000 additional farms on which farmers rent land and have established acreage bases eligible for payments in the transition program.

This gives Wisconsin a total of 93,000 farms with proven base acreages.

Caruso gave an idea of how massive this payment program will be when he noted that all 60 county Farm Service offices — working together — need to sign up 2,500 farmers each working day until the final July 12 date.

It is a major job. Not only does it involve the farms of 39,000 state farmers, but the system of payments also is based on rented crop program base acreages where renting farmers get the payment. That makes for a total of 93,000 Wisconsin farms with active base acres.

Much of the work at county offices is done rapidly on computers and farmers will leave knowing how much they will be paid in the seven-year program. An advance payment scheduled to be made within 30 days of their office visit also is available.

Many offices have made the sign-up system easier by scheduling office times for farmers to get to the office and complete the sign-up requirement.

And all eligible farmers have been advised that they have between now and July 12 when the sign-up ends. If they miss that requirement they will not share in any of the financial aid.

Annual payments in Wisconsin, if all of the 39,000 eligible farmers sign up, will range from $89.2 million to $162.8 million, until the special phase-out program ends.

This payment program is one of the largest Caruso has handled since he became state executive director of the state FSA in 1993.

Dairy Veteran McKinley Called Upon

October 6, 1996

When some of the nation's top dairy leaders were in Madison this summer, they had a chance to see just how thorough Myron R. McKinley is in handling major dairy assignments for the dairy division of the U.S. Department of Agriculture.

Now McKinley, 39-year veteran market administrator of the Chicago Regional Milk Marketing Order and other markets, is making plans to be back in Wisconsin for the state's third dairy reform session on Wednesday in Milwaukee at the Manchester East Hotel, Port Washington Road, from 9 a.m. to 12:30 p.m.

This seminar is where Richard McKee, director of the Dairy Division of the Agricultural Marketing Service in Washington and the man responsible for the nation's entire Federal Milk Marketing Order Program, will give his review and update of the massive reform process, which Congress triggered in the 1996 Farm Bill. The federal order provisions will dramatically change in the next three years.

It is getting to be crunch time for McKinley and three other marketing order administrators who are responsible for drawing up the recommendations for changes that lie

Myron McKinley, right, gets a Packers memento

ahead for issues like basic formula milk pricing, the committee of which McKinley is chairman; pricing structure, organized by Arnold Stallings; and milk classification, by James Williamson.

All this professional leadership will come up with programs and changes to be part of the new federal order system, which reduces the current 33 market orders so there will not be more than 14 or less than 10 to serve the national market.

This Milwaukee program was scheduled by the Middleton-based Wisconsin Dairy Products Association.

The basic formula price is the base on which federal order milk pricing is made for nearly all of the nation's dairy farmers. California uses a state market order for its dairy pricing, but has been invited to join the new federal order program to complete national coverage.

McKinley is regarded as the dean of federal order administrators and has a major role in determining pricing provisions offered in the reform program.

The pricing system being used now gave dairy farmers a new record price for base milk supplies since the old Minnesota-Wisconsin milk price set the previous record price level in December 1989.

The Milwaukee session also will conclude with professional responses from dairy cooperative spokesman Paul Christ of Land O'Lakes, Arden Hills, Minn., and proprietary dairy plant views from Ted Galloway, of Galloway Co., now Waterford Food Products, Fond du lac.

McKinley in his 39 years of service with federal milk marketing orders has worked as administrator with the Eastern Ohio and Western Pennsylvania market order and has been administrator with the Chicago Regional Marketing Order, which covers much of Wisconsin, for the past 13 years.

He was a farmer for seven years and was employed later at producers' organizations in Dodge City, Kan., and in Akron, Ohio. During World War II, he served in the Navy.

As a result of his many experiences with dairy plants and producers, McKinley has frequently been called to Washington by the USDA dairy division for discussions on major issues facing administrators of federal orders.

Ideas for Milk Pricing Reform Collected on Computer Disk

November 24, 1996

Dairy farmers with a computer will have a chance to fully learn what is being considered in the major overhaul of Federal Milk Marketing Orders and milk pricing.

Richard McKee, director of the Dairy Division of the U.S. Department of Agriculture, and four market order administrators — each heading a special study committee — are working on ways to reduce the number of milk marketing orders and rewrite the basic formula price, which determines the monthly milk price for farmers.

McKee most of all wants to learn what dairy farmers and milk handlers want this reform to accomplish.

When starting the revision last May, he asked dairying's "interested persons" to offer their ideas on how to improve federal orders and its pricing system.

Thus far, McKee has received 125 letters outlining changes farm and industry people want to see incorporated in the dairy reform.

The dairy reform initiative was ordered in the 1996 Farm Bill.

Secretary of Agriculture Dan Glickman has emphasized that the reforms will be ready for Congress well before the 1999 deadline.

McKee has offered a computer application, which gives dairy farmers and processors a chance to find out what is being proposed by those who have responded to the request for ideas.

He is offering a computer compact disk, which contains all 125 documents received thus far.

And McKee said he believes that knowing what has been recommended thus far will stimulate others to submit their ideas.

The disk is available at the cost of $65 each, and the cost can be a tax-deductible business expense.

There also is a shipping and handling charge of $3. Overnight shipment is $15.

To order the compact disc, call Tina Batcheller, of Input Solutions, at (800) 935-7226.

The Dairy Division of the USDA also will cover costs of using a Help Line at Input Solutions. Such help will be free and will be available all hours.

U.S. Dairy Market News Office is One of a Kind

November 19, 1989

Markets and more markets — every blessed working day.

Markets may seem a little like dullsville, but not so at Don Nelson's, Dairy Market News office, on Badger Road on Madison's South Side.

Nelson and his four regional dairy reporters, with a combined total of 63 years in the business, runs the federal government's only dairy market news office in the nation.

Consequently, Nelson's crew of specialists is the prime source of news for the nation's dairy markets.

And that dairy market is hot.

Right now there isn't a single farm commodity that's seeing more demand and market action at home and around the world than cheese, nonfat dry milk, fluid milk and whey products.

"The reason for all the market activity is that dairying is not a support market propped by the government. It's a free market with volatility and the need for a great amount of information," Nelson said.

Madison's office, where Nelson is national supervisor, is an agency of the U.S. Department of Agriculture's Agricultural Marketing Service and its Market Information Branch in Washington.

Nelson grew up on a dairy farm near New Ulm, Minn. He milked Holsteins and showed calves and cows during his 11 years as a 4-Her.

He is a graduate of St. Olaf's College and holds a masters degree in agricultural economics from the University of Minnesota. He has been with the marketing service for 14 years.

He and his wife, Krista and two sons, moved to Madison last year, when veteran Tom Hadfield, Waunakee, finished his dairy marketing career.

Nelson is reporter, editor and publisher of the weekly publication, Dairy Market News, which has 1,900 U.S. and foreign subscribers at $30 a year.

Don Nelson

The mission of the office is to disseminate dairy news. But first they have to gather it from telephone contacts in five regions of the country.

Reports from dairy plants, processors and other dairy sources, are compiled in weekly reports for each region, on prices, demand, and supplies of milk and dairy products.

Nelson's office is the government's only national dairy news office. The office also logs calls from Australia, New Zealand, countries of the European Community, Canada and Mexico.

The office has a battery of six code-a-phones answering 600 calls weekly. One week last month the office logged 350 call for dairy information.

If this office doesn't know what's happening in the dairy world, it hasn't happened.

For instance, when world demand was at its peak for nonfat dry milk, Nelson and his crew learned millions of pounds of U.S. powdered milk was stored in West Germany and was being shipped back to the U.S. It is not an import, but it is labeled as "returned goods" in an arrangement handled by U.S. Customs Service.

Nelson said the government is delaying dairy product shipments to Mexico for 90 days to build up supplies for the domestic market over the holidays.

"I think the dairy market over the next few years is going to be more volatile than before. That's because some government programs helped get rid of cows and surplus dairy stocks," he said.

"With these price signals there is the possibility production may start to increase across more areas of the country. If that happens there is a good chance of going back into a surplus on some of our commodities," he added.

Retirng ASCS Officer Looks Back on his Career

October 15, 1989

When Milford Thompson retired this month from the Wisconsin Agricultural Stabilization and Conservation Service office in Madison, a 37-year chapter of Wisconsin dairy history closed.

Robert Van Liere, of Salem, a former Pure Milk Association officer who served with Thompson as a PMA director said, "Thompson was a leader who never sat back and just talked about what could be done. Instead he was a positive thinker who could clearly explain the most complicated issue and get the job done."

Furthermore, Thompson put his energies to work in both dairy cooperative and public leadership roles to help nearly every dairy farmer, not only in Wisconsin, but in the nation, Van Liere said.

Thompson was 28 when he was first elected president of the Mount Horeb local of the old Pure Milk Association, a cooperative with a large Wisconsin membership. Later he was elected a district director.

Thompson is 65 years old now and said he is looking forward to enjoying the change in his lifestyle that he says will include travel and some warm winters.

He started helping dairy farmers in dairy policy development through PMA. Van Liere said he provided the leadership to achieve what farmers in his district wanted most — to have dairy cooperatives work together to improve farm milk prices.

This move was underway in the late 1950s when dairying faced tough times and some farmers in Thompson's district openly were working for a strike as a way to gain milk price power.

But Thompson was able to aim for market power based on the strength and power, which was used rarely by dairy cooperatives. Farmers in the district wanted such a plan and advanced it with a favorable district vote at a meeting in Evansville, which became known as the "Evansville Expression."

The plan called for their representatives to negotiate an increase in milk prices with dairies in the big Chicago Regional Milk Marketing Area.

Van Liere said this led to negoitated premium, first known as the superpool premium, of 6 cents per 100 pounds for Class One milk. The premium now called a cooperative price is more than $1 per 100 pounds on 20 percent of the Chicago market's milk going into the bottle. Cooperative price action as evisioned by District 19 farmers started about 30 years ago.

Early premiums totaled about $1 million a year and sometimes rose to $7 or $9 million. For a time premiums were the source of dairy promotion funds in the Chicago market.

Eventually nearly every federal milk market order area in the country — those controlling 85 percent of the milk supply — used premium pricing for the benefit of farmers. Thompson has seen the District 19 plan add more and more millions of dollars to dairy farm income around the nation.

Van Liere said the District 19 plan was successful

because dairy firms accepted the idea to ensure their milk supply. At the same time they were confident all dairies were paying the same price.

In 1969 Thompson made a major career change and received a state ASCS appointment as a district director responsible for review of farm program administration that eventually involved 40 Wisconsin counties.

On the basis of his efficiency, he was promoted to production adjustment and compliance specialist, which made him responsible for the dairy diversion program for 13,400 state dairy farmers in 1984 and 1985 to reduce milk production.

Then he was in charge of the whole-herd dairy buyout program used by 1,650 dairy farmers in Wisconsin in 1986 and 1987 when 108,000 dairy animals were cut from state herds to reduce milk production.

Milford Thompson

Thompson said cost of the buyout in the state was estimated at $125 million, but all costs were erased by dairy farm assessments, which matched total federal costs in the state.

"Taxpayers didn't spend a cent on that program in Wisconsin although the total cost nationally was $1.8 billion," he said.

There never was a dull moment for him in ASCS state office since he also was responsible for the feed grain program designed by Congress to limit grain production.

Through it all, Thompson hasn't changed.

He still has that youthful vigor, the same easy friendly smile and, most importantly, the same valuable farmer logic on how to get a job finished for the benefit of farmers.

Chapter 8

Education, Research and Extension

They're the Big Cheese of State Dairy Research

January 10, 1993

UW cheese researchers Mark E. Johnson and Walter V. Price.

Two masters of cheese research at the University of Wisconsin — a generation apart — stood shoulder to shoulder to celebrate their mutual achievements recently.

Professor Emeritus Walter V. Price, 96, the master of cheese and the first Pfizer-award winner in 1959, joined senior scientist Mark E. Johnson, 41, the 1992 and current Pfizer award winner.

They recalled their work in cheese development laboratories on campus where they had a chance to make major contributions to Wisconsin's famous cheese industry.

Price came to Wisconsin after World War I from Cornell University and became the master of cheese product development, particularly for the improvement of Cheddar and brick cheeses.

Johnson worked his way through college making Cheddar and colby cheese while attending South Dakota State University.

He completed his doctorate at South Carolina State University in 1980 and has been on the Wisconsin faculty since then to become the recognized researcher of the UW's Center for Dairy Research for his development of Wisconsin-style Havarti cheese.

Price has nearly complete recall of those days in Hiram Smith Hall laboratory and in Babcock Hall, where he established himself as one of the pioneers of research to improve cheese quality.

But there is a difference.

To work out the mathematics to perfect his formulas for Cheddar and brick cheeses, Price used the pencil and long multiplication.

Johnson, on the other hand, uses a computer for his cheese comparisons and takes five minutes to do what Price labored over for half a day.

Price was a pioneer in heat treatment of milk that he started as a graduate assistant in the dairy school at Cornell.

They still joke with Price on his knowledge of milk heat treatment, for after all he received his basic training in heat as a Naval fireman who stocked the boilers with coal aboard the USS Maine, the Naval battleship built to replace the ship destroyed in Cuba's Havana harbor.

And his experience at Cornell and Johnson's first experi-

Mark Johnson and Walter Price

ence at Wisconsin were almost parallel.

When Price showed up for work at the Cornell laboratory, his professor promptly left for a one-year sabbatical and left him in charge of cheese research.

Johnson also showed up on April 22, 1980. Norman Olson, dairy center director, showed him the laboratory and promptly left for conferences in Europe and the Soviet Union.

So both Price and Johnson started their best and most notable research when left to themselves on their initial research appointments.

The Havarti development was funded by Wisconsin dairy farmers through the Wisconsin Milk Marketing Board. Both Johnson and the Milk Marketing Board had a winner because all Havarti cheese was imported from Denmark at the time.

"I was happy to get the Havarti research because it was one of the cheeses that surveys indicated would be one of the up-and-coming cheese varieties.

But Johnson had to start from scratch because he had to develop the formula to make a cheese comparable to Havarti.

"Getting the formula was easy and the cheese matched the Havarti composition. But it was more difficult to match the flavor or the cheese body.

"Really, it was similar to the work Dr. Price did on his brick cheese, since I used a washed curd method used in brick cheese to control the acidity and flavor, too," Johnson continued.

His challenge was to make a tasty cheese that would stand on its own and last for three months without getting soft. He wanted a cheese with a longer shelf life.

As Johnson told his Havarti story, Price gave him credit for his work.

"It seems to me," Price said, "that what Mark is doing is the advance stage of all the things that interested me in improving cheese."

Johnson is particularly interested in the development of a low-fat but full-flavored cheese.

"Now we have developed cheese with a stronger flavor, while getting rid of the fat. The cheese is still being tested in the laboratory, but it consistently has a good flavor and quality," Johnson said.

Agriculture Links Father, Son

January 30, 1994

There is a professorial link for the Grummers of UW-Madison:

Ric Grummer, 38, associate professor, in dairy cattle nutrition research (60 percent) and teacher (40 percent) since 1984, in the department of dairy science, and his father, Robert Grummer, 79, emeritus professor of meat and animal science, since 1941, a nationally recognized leader in swine nutrition research, and pork quality breeding on the UW-staff since 1941.

The academic and professional link of the father and son was nutrition research.

Robert (better known as Bob) in swine nutrition, and Ric, in nutrition research for dairy cattle.

For each of the Grummers, there was a chance moment of opportunity that set the path to their university futures.

Bob was an Iowa farmer's son from Benton County, where dairy dominated, on the advice of his high school teacher, was headed to Iowa State (College) to become an engineer in 1936.

The farm depression still had its impact in 1936, but he headed for ISC with $36 in his pocket and a desire either to be a farmer or an engineer.

Ric and Robert Grummer

He hitchhiked to the Ames campus with a friend who was to register in animal husbandry.

For a moment the two paused to decide, with the flip of a coin, where the first stop would be.

Agriculture got the flip and Bob became a winner with a future.

While waiting for his friend, he chatted with an ag professor and soon Bob accepted a full scholarship, saving his day and making his future.

After graduation, Bob told his professor he wanted to work and then start graduate school. The professor remembered and when UW's Gustav Bohsted called him seeking help, Bob, by then a county 4-H agent, was the chosen candidate.

He took charge of all UW campus livestock plus the meat lab. He also started graduate school and while working earned his advanced degrees.

Ric also had his moment. He enrolled at the UW-College of Agricultural and Life Science and studied to get into the veterinary school. He took a job handling campus dairy cattle.

While in the barn one day, Professor Larry Satter

stopped by and asked Ric to considered working in his nutrition laboratory.

His new focus became cattle nutrition at Madison and he received his advance degrees at the University of Illiinois, where he also met Mary, — a nutritionist — who would become his wife.

His chance to be on the UW dairy science faculty was perfectly timed because there were two faculty openings when he applied in 1984 — along with 39 other hopefuls — and was accepted. He knows his dairy farm work too, because he was an intern earlier as an undergraduate on the Les and Peg Ballweg farm in Dane County's town of Roxbury.

Like his father, Ric spends volunteer time with Extension speaking to professional and farm groups on nutrition. He is eager to have farmers and feed professionals share the results of UW nutrition research.

Retired, Bob is never without things to do. Before the death of his wife, Lee, they started restoring old furniture and other art works. They held an annual sale to reduce the inventory. Many of their customers were young couples furnishing their new homes.

Earlier this week, Chef Bob prepared a dinner for 10 guests who feasted on his perfect loin roast and finished with a serving of his own hickory nut pie.

Both were good, but Bob set two pieces in the refrigerator and later noticed that only one piece of hickory pie remained.

A grateful guest confessed that he just had to have another piece of pie because "I never tasted anything like that."

Bob will be honored March 4 in Kansas City, where the National Pork Producers Council will present him with the organization's Distinguished Service Award.

Bob was nominated by the past presidents of the council and confirmed by the groups executive committee.

Graf's Latest Takes him Through Eastern Europe

October 31, 1993

After 33 years handling Agricultural Economics classes and Extension sessions for UW-Madison, Emeritus Prof. Truman Graf hasn't slowed down.

As expected, he has stayed current in his analysis and reading of dairy plant economics and business marketing .

Graf had to be current in economics because he just completed 14 months of providing technical assistance at dairy plants in Bulgaria, Hungary and Poland. In the Ukraine and Russia, he and a faculty colleague, William Dobson taught classes.

The overall goal was to help adult students fully understand privatization of business and how they can gain the advantage in a free-market economy.

Graf's classrooms and technical assistance sites had the kind of students every professor admires — interested, eager to learn, curious and willing to discuss the issues.

Graf said the business issues were so new to students, the course material had to start with the basics to make certain his class members knew what was going on and how it could be applied in the emerging business world. He worked with a translater, and all of the projected overhead charts he used were translated for the class. Students also received Graf's final summaries and recommendations in their own language.

And when students wrote their course evaluations, they gave "trainer" Graf exceptional praise and good marks for the help he had given them.

Graf has that touch, having grown up on a dairy farm in New Holstein. He started milking cows at 4 years of age because he was eager to do the job.

He was a Navy dive bomber pilot in World War II and received his bachelor's degree, master's and doctorate at UW-Madison, where he joined the faculty in 1952.

Truman Graf

The men and women of Eastern Europe are in a tough period of business transition and new economic thought, Graf said, because in the former Russian-dominated totalitarian system they were told what to do.

Now they have to make the marketing and operational decisions to keep their new private businesses going. He worked with new dairy plants that, in some cases, were marginally profitable, but had managers who were "very energetic and innovative," Graf said.

However, they were reluctant to bring new partners into the business to gain additional capital, preferring instead to continue with new programs with modest expenditures.

One of the exciting products at the plant in Karabunar, Bulgaria, was feta cheese, a Greek cheese made with sheep milk, which has gained strong consumer demand. It is regularly used on salads.

Graf was in Bulgaria, Hungary and Poland as a "trainer" in a joint project of Land O'Lakes, the giant dairy and food cooperative, and the U.S. Agency for International Development. In the Ukraine and Russia, he was involved in another joint project with the agency and Iowa State University.

"These projects make two points," he said. "Anything the U.S. can do to help Eastern Europe will be very important to the U.S. and the world generally. And if this contact can improve business operations, income and living conditions, it will be a big plus for peace and stability."

A Look into Farm Co-Op Future

March 17, 1996

Wisconsin is home to scores of agricultural cooperatives, but on April 1 it will take a look at a new generation of the farm cooperative future.

Robert Cropp, dairy economist and a familiar figure in Wisconsin dairy affairs, is wearing his other hat — as director of the University of Wisconsin Center for Cooperatives. Cropp's center has teamed up with several organizations for a one-day program to inform state farmers of what's involved in the "New Generation of Farmer Cooperatives."

Robert Cropp

The April 1 program at Stevens Point, is based on the experiences of farmers who have had the courage to carry out new ideas for closed cooperatives, where members buy a share of stock with their production — such as a particular crop, milk, beef or hogs.

Cropp has assembled representatives of a diverse group of new cooperative leaders. This "new generation" involves more than 50 newly organized farmer cooperatives, which now have a record of revitalizing rural areas of North Dakota and Minnesota.

One interesting change is that the plan, which farmers create to add to their income, will only take as much out of a member's production as it takes to meet demand for plant products, Cropp explained.

For example, out of the supply members ship to their plant, there may be an assessment to pay for the plant they decided to build.

At a milk plant, for instance, the participating farmer may be paying $1.08 a hundredweight out of his milk check. But it is paying off the investment, which is performing the service members need.

Because of the logic the new cooperative plan is built on, cooperatives have been able to gain bank financing for plants and thus have additional sources of financing to work with.

The theory behind the new cooperatives is similar to that behind Ocean Spray, which has developed a major industry and respectable returns for Wisconsin cranberry growers.

According to Cropp, cooperatives can succeed when ideas their farmer members have carried around for years on how to do business together help improve member income and develop opportunities for employment in the surrounding community.

"Each of these new cooperatives have added value to the commodities produced and marketed by their members," too, he said.

"More than this, these cooperatives provide much-needed new jobs to their respective rural communities and help retain the business profits generated within the community," Cropp continued.

"For the Stevens Point program speakers will explain the first development that came with new generation cooperatives like the Wisconsin Specialty Cheese Institute and the Coulee Region Organic Producers Pool (CROPP)," he explained.

Computers Leap Language Barrier

June 17, 1990

Even the computer can be a factor in international programs to help people improve their lives. Allen Wollenzien, who trains people in county and state extension offices to use computers, is developing a computer system overseas and he is convinced it works.

He is with the WISPLAN computer program of Wisconsin Cooperative Extension on the UW-Madison campus.

Wollenzien, 51, who lives in Stoughton with his wife, Gail, and family, came from Sheboygan County to join the Dane County Extension staff in 1981. He served until 1988 as a county dairy and livestock agent.

But while studying for his master's degree at the UW-Madison he linked up with computers.

He took a liking to computers and when the computer training position opened he was experienced and ready for the new professional opportunity Extension offered in its computer world.

But Wollenzien wears two hats — both of them helping people.

He spends most of the time involved with computers used for agricultural programs that help Wisconsin farmers with livestock feeding, record-keeping and decision-making. He also trains office workers for use of the computer in printing extension information bulletins and setting up a records system.

During his vacation and leave time, he devotes his energy to a Stoughton-based organization: Thoughts of

Faith.

This group is developing programs to help the Ukrainian people make the transition from a totalitarian regime, to move more quickly into a free economic and political society based on principles of democracy.

Last fall, program director Pastor John Shep purchased two computers, which he and Wollenzien took to the Ukraine.

"This all sounds great," Wollenzien said, "but how am I going to teach these people to use computers when they don't speak English and I don't speak Ukrainian."

"I guess my prayer was answered the day we set up the computers in a school in Lasav, located in the Western Ukraine, which is the heart of wheat growing country," he said.

Just as the computers were being set up two representatives from an area technical institute arrived at the school. "We hooked up the computers, got them fired up and the minute we hit the keyboard and started using the DOS (the disc operating system computer language) that took care of the communications problem.

"Both of them knew DOS and I know DOS. By just typing and shaking our heads a lot we communicated in the common computer language. We communicated what we wanted to do with those computers," he continued.

Wollenzien was in the Ukraine for two weeks in September and went back in late October to check out the system and program performance.

The system is working so well 500 more computers have been sent to the Ukraine students to use. These personal computers were all purchased with donations to Thoughts of Faith.

A chance to learn English is still the principal request of Ukrainians, Wollenzien said, and the computers have the software to teach English. For instance, the letters of an English word to be learned are scrambled on the bottom of the screen.

When the letters are correctly put together on the top of the screen, the screen explodes into a picture of what the word describes.

"The kids sit in front of a machine, go right after it and start learning," he said.

The computer training plan is working so well, Wollenzien said, that on the second visit he was asked if someone could come over to the technical institute and teach basic agriculture "as you know it and do it in America."

Agricultural technology and farm management practices are something these people have lost in the last 40 years, even though they live in a great agricultural region, he added.

"We are expecting Shep to come back to Stoughton from the Ukraine very soon with a formal letter requesting that kind of farm training program," Wollenzien said.

"Maybe we in UW-Extension can get involved in some kind of an international extension project. This may be possible through the international program on the agricultural campus," he said.

Wollenzien met Shep when he was pastor at the West Koshkonong Church in eastern Dane County. Shep's parents left the Ukraine early in World War II and Shep was born in a refugee camp in Austria.

Later, after serving at West Koshkonong, he founded Thoughts of Faith. Shep, who kept in touch with events in the Ukraine, knew far in advance that changes in the economic and political systems in the Soviet Union were coming.

So when President Gorbachev introduced the glasnost and peristroika concepts for change Shep was ready to move and start helping the Ukrainians.

"What farmers are saying now is, sell us the land or give us a loan program so we can acquire it. This is major part of the whole peristroika program for economic change in the agricultural area," he said.

"We have to remember that this period of economic change also involves businesses and their part of the economy. It could start with agricultural cooperatives as collective farms shift to private operations," he said.

"The window is open for change. Hopefully it stays open so we can help," Wollenzien said.

Dairyman Faces Soviet Challenges

February 11, 1990

UW-Extension Dairyman Terry Howard has had some big challenges in his life, but he met them with his 6-foot 7-inch frame and many times came out a winner.

Terry Howard

For instance, Howard was the highly recruited center on the University of Nebraska basketball team and had to play against 7-plus footer Wilt (The Stilt) Chamberlain of Kansas State. The Cornhuskers won, 43-42, just two weeks after K-State had demolished them.

But that comeback against the nation's top team has given Howard a philosophy showing there's always a chance to come out a winner.

He is applying that comeback philosophy now internationally as he works periodically with a team of dairy cattle experts in seminars, by making dairy farm visits in Poland and the Soviet Union, and as he follows dairy activities in Hungary.

He was on two such trips last year as the drama of

changing governmental systems surged through the Soviet Union and Eastern Europe.

Howard was with the president of the Soviet Republic north of Leningrad who was host of the dairy team when he declared that "under my administration producing more food is my number one priority and problem."

And he was with Soviet school children in a city that housed the only concentration camp for children during World War II. He joined them in singing the international theme song of freedom for the people, "We shall overcome."

And he was with people who said: "we are almost afraid to hope we can change." But they know they must, he added, because the production and distribution system has broken down and people have no incentive to do a better job.

In Leningrad there were great lines outside every jewelry store and people explained to Howard they were converting rubles to gold because of the rumor that currency values would change.

The official exchange rate is 1.6 dollars to the ruble, but out on the street a dollar commanded 12 to 15 rubles.

One of the great Soviet shortages is toilet paper. You can go into public buildings and find torn-up newspaper for toilet paper, he reported.

"The cat's out of the bag and people are seeing the Western World on TV and how people can come and go as they wish," he continued.

The longest lines he saw in Poland were at used book stores. At restaurants there was one vegetable — green peas — with lots of potatoes and limited meat servings.

In the Soviet province he visited, the country looks much as it does in the Duluth-Superior area. The dairy emphasis is to boost milk production from an average of 5,500 pounds per cow per year. That's about a third of what it is in Wisconsin.

The province wants to buy bull semen to improve the herd, buy embryos to get the best cattle quickly, and buy the expertise to do the job right.

It is going to take hard currency (dollars) to do it, but they are determined to generate the hard currency, he said. "In dairying they are where we were 30 years ago. The U.S. has the dairy technology they need. There is a tremendous potential market for Wisconsin firms as a result," he asserted.

Howard, who grew up on a dairy farm in Colorado with Guernsey cattle, didn't go into dairy farming because the 1930s Depression and brucellosis led to the loss of the family's farm.

His father, George, was a professional cattleman and managed herds in Illinois, where Terry went to high school, and later in Nebraska.

"At least half of what I know in dairying came from my father because he could get more milk out of cows than any man I have ever known. It used to bug me that he could get more milk out of cows than I could. I was well blessed because my father would talk to me on what we were doing with the daily herd." he said.

Howard managed the dairy herd at the University of Nebraska and was a dairy instructor for five years. He went to Purdue University for his masters and doctors degrees.

He came to Wisconsin in 1966, established himself as an extension dairyman and then extended his work in farm computer specialties to aid farmers in cattle ration balancing and numerous other farm decisions. He was the first director of WisPlan, a network linking all extension office computers to the UW-Madison campus.

Farm Roots Nourished Professor

July 17, 1994

ITHACA, N.Y. — Just how a couple of 120-acre farms near Granton in Wisconsin's Clark County can affect the professional career of a teen-ager in South Milwaukee is the story of Cornell University Professor Andrew M. Novakovic.

As a youth, farming impressed Novakovic, 54, who today is chairman of Cornell University's Department of Agricultural, Resource, and Managerial Economics, which many know as ag econ.

His mother, Esther Schinslog, a nurse with a doctorate, was born and raised near Granton in Clark County, where her father and brother had farms that attracted Novakovic.

Novakovic's father was a Serbian from Yugoslavia, who had served in the Yugoslavian Army in World War II, was captured by the Germans and spent five years as a prisoner of war.

When liberated by the Americans, a Serbian-American sponsoring group helped him get settled in South Milwaukee, get a job, learn English and have a chance to meet Esther Schinslog, a cousin of Middleton's Bill Schinslog.

When Navakovic was in high school, he pressed his parents to let him spend his summers on the Clark County farms. He came away knowing Wisconsin agriculture — dairying, haying, chores and just plain everyday work and the sweat it triggers during the 90-degree summers.

"I just loved the farms and the work. The farms were small with herds of 25 to 30 cows each and certainly not on the cutting edge of technology, but I learned," he said last week at his office in Cornell's Warren Hall.

Andrew Novakovic

"Farming was a gripper for me. It ended up by shaping my career," he said.

When he was a high school senior he took a course in economics and liked that, as he did agriculture and farming. He enrolled at the UW-River Falls, and decided early he wanted to go on to graduate studies and to couple his two main interests by studing agricultural economics.

Novakovic wanted to join the ag econ staffs at either Wisconsin or Minnesota, so he took his master's and doctorate studies at Purdue.

Neither school was hiring at the time he finished and after an interview with Cornell University he was offered a position.

His wife, Martha, whom he met at River Falls, helped him learn about Cornell. She was a resident of Cornell, Wis., which was planned by Ezra Cornell, founder of Cornell University. Thousands of acres of Wisconsin timber were deeded to Cornell University as part of a federal land grant because there weren't enough federal holdings in New York State.

Novakovic was asked to join the staff in Ithaca in January of 1978.

He fit right into the Cornell ag econ plan because his research and teaching of dairy marketing fit the strong program at Cornell.

Novakovic said the issue in the U.S. dairy industry today is where it is going.

He said since 1988 there really have been no major dairy price supports, except for butter. All parts of the industry are learning to live without that federal mechanism. Both the small and large processors are worried about the changes.

"I find it quite ironic today to be talking to people who know what agriculture was like before 1930, when there were no federal (support) programs," he said.

This situation gives a new challenge to Novakovic and his economists, whose job it is to determine what kind of a dairy industry the nation will have.

He said that not all conditions today in dairying are "terribly different than they were 100 years ago" — because there is no food price inflation.

Novokovic has been in Wisconsin frequently as a dairy speaker. He also spent a sabbatical year in Madison and Middleton in 1986 and 1987 working on dairy products consumption data with the Wisconsin Milk Marketing Board and as a researcher and as a visiting professor in the UW-Madison Agricultural Economics Department.

Faber Herd Gets Heat Test

February 4, 1996

ITHACA — An exceptional herd of Wisconsin dairy cattle has been moved to another state in the name of research.

Steve Faber, who developed a Brown Swiss herd that is one of the three top milk production herds in the nation, relocated his 150-cow milking herd from Ithaca, in Richland County, to the University of Arizona in Tucson.

The massive trucking move of the herd started July 23 and the last load of cows was moved on Dec. 27. Steve, 47, is in Arizona now to manage and do some instruction for the Arizona University Dairy Research Center, where his Brown Swiss — leased to the university — will be paired, side by side, with a 150-cow Holstein herd with similar milk production records.

He is not alone in meeting the responsibility. His wife, Nancy, formerly of Sextonville, does instruction work and computer work on cattle, including taking care of registrations for the two herds.

The research is focused on how cattle can maintain milk production despite the effects of Arizona's heat. The research is being conducted to determine which breed, under the same management system, can take the heat and show it "is more capable of producing the same amount of milk, consistently."

Brown Swiss are more tolerant of heat than other breeds, despite their origin in the mountains of Switzerland. It will be an interesting match with the Holsteins, since it is a major factor in cattle exports to Southern areas.

But it is important to the Swiss breed, too, as Howard Voegeli, Monticello, Brown Swiss breeder, noted: "We have a lot riding on Steve's findings."

Professor Roy Ax was on the UW-Madison Dairy Science staff before his appointment as head of the UA Department of Animal Science.

While in Madison, Ax got well-acquainted with the exceptional talents of Steve Farber, who fully developed the potential of the Farber herd through concentrated and extended study of rapidly changing agriculture and dairy management.

Ax, at Arizona, led development of the Arizona Dairy Center for this critical research on heat tolerance and milk production.

Actually, this whole Arizona shift of Faber Brown Swiss to Tucson is a wonderful family story for the Fabers. The herd development story started after World War II with Steve's father and mother, Norman and Marie, in Richland County, a short distance east of Ithaca.

Later his Brown Swiss operation was based on the farm

along Highway 58 on Ithaca's north side, which grew and prospered.

But Norman, a natural innovator, planner and designer, credits Steve with full development of the Brown Swiss potential, which steadily built the 140-cow herd from 21,000 pounds of milk per cow to a high of 27,000 pounds per cow.

He gave Steve a father's ultimate assessment: "Steve did everything that was right. And he never ceased to learn. He went into studies of every phase of dairy agriculture as deeply as he could."

Russian Livestock Geneticist Visiting Wisconsin Farms

February 5, 1995

Madison has a special visitor whose goal is to gain experience and knowledge of animal genetics and research. She also plans to study the business world, to demonstrate that she can contribute much to major animal programs when she returns to research work in her native Russia.

Galina Petrovna Simonova, a geneticist and researcher at the All Russian Research Center of Sheep and Goats, has the special distinction of being the first visiting scientist brought here for extensive study sponsored by the International Chamber of Agriculture.

The chamber is the international unit organized by Madison entrepreneur William T. Graham and is dedicated to improving the quality of life around the world through agriculture.

It is part of the World Dairy Center being developed near the new state Department of Agriculture building. Simonova, who has been here for about half of her six-month Wisconsin stay, is the recipient of a Special

Galina Simonova and Bob Walton

American Business Training program grant from the U.S. Department of Commerce.

She speaks and writes English exceptionally well — an accomplishment she mastered by attending an agriculture school where Russians learn English and prepare for assignments in English-speaking countries.

She lives at Stravropol, which is 800 miles south of Moscow and about 300 miles from the beleagured enclave of Chechnya.

She is married to a teacher, Alexandr, in Stravropol, and they have a 9-year-old son, also named Alexandr. Her husband is a Russian decathlon champion and teaches the sport.

During her stay here, Simonova also is expanding her knowledge of the world by learning about Americans. Before she came here, she held the belief of many Russians that Americans were the one group of people in the world who had no problems.

"But when I arrived here and met many families, I really understand that American people also have problems," she said.

But, she added, "American people also work hard to solve their problems."

The American experiences she has had in the past three and a half months have kept her exceptionally busy, but she has found time to absorb a fair share of the culture.

"I like the choices Americans have and how they show the world that it can happen to anyone who wants to work for it," she said.

"I also like the freedom people in America have," she added.

Her hosts are Robert and Jan Walton, who live on a farm south of DeForest and have a Simmental beef herd. Walton is past executive officer of ABS. Galina helps the Waltons with the beef cattle and has shown she can handle a manure fork and drive a tractor.

That's because she lived on a Russian collective farm. Her official visits include talking with a number of Wisconsin goat breeders, especially those who have organized goat milk distribution programs.

She said she was thrilled with the trip to the Sunshine Farm of Dan Considine, near Portage, where there are 1,600 goats, 200 of them milking.

She soon discovered that the Considine family ranks as leaders of the goat industry in Wisconsin and the United States.

Her American goat farm visits also included a tour of the Ken Kromrie Goat Milk Dairy near DeForest. Kromrie has an operation with 400 goats. Because of poor production, goat milking is limited in Russia, but she would like to see milk breeds improved through better breeding and that is one of her goals when she returns home.

She also explored the field of business and agricultural

banking with Debra Linz, founder of the specialized bank in Sauk City.

Simonova commented on how banker Linz was "another woman who is an entrepreneurial model." She also is getting the lesson in business from Marci Hein of the University of Wisconsin Foundation and Sue Springman of the Munz Corp. One of the requirements of her work back in Russia with the Sheep and Goat Institute, she said, is that staff members must have a direct tie with a farm, especially one with goats and sheep.

So when Simonova refers to "her farm," she is really explaining her professional link with a collective farm, a large farm organization which still exists in Russia even with the change in economic theory and Communist farm ideas.

The farm is where her father and two brothers work. It is at Svetlograd, where she was born and her family lives. Her father is responsible for a flock of 1,200 sheep. The farm also has 8,000 goats.

Simonova explained that some of the goats are breeds kept for shearings of fine fleece or down. Many of the goats are used to provide meat.

The purpose of her trip to Madison was to improve her knowledge in advanced studies of animal genetics with Dave Thomas, UW sheep specialist.

With this training she will be able to help the people of Russia when she returns to her research post.

Dairy Expert Stays Current and Active During Retirement

July 21, 1996

When it comes to the years that really count, Evert and Ilma Wallenfeldt are living comfortably in their apartment home at Oakwood Village, a well-known retirement center in Madison.

Evert, 92, has a way of keeping in contact with friends and events of interest, just as he did when he retired as an emeritus professor of dairy science and extension at UW-Madison 26 years ago.

He keeps in touch with his chosen field — dairy — and his many friends in an organized way with two monthly meetings.

One of the gatherings has a dairy emphasis and a speaker to enlighten and challenge members. The other session includes his retired agricultural friends at a breakfast session with no speakers, but lots of conversation across a long table and an exchange of many good natured opinions.

Evert still has a following because of a career in dairy science that helped Wisconsin gain its reputation for quality

Evert and Ilma Wallenfeldt

milk production. But to this day, Evert is a man who never takes credit for what he contributed to Wisconsin and the nation's dairying, because — in his words — "It was those who made the dairy plant operational decisions to install the system achieving milk quality, who really did the job."

The marvel of Evert's influence, however, was that he did more than convince dairy plant owners and operators to emphasize quality milk production. He also encouraged them to maintain high standards, by incorporating test laboratories in their plants to ensure the continuing quality of milk.

The Wallenfeldt story is a Wisconsin story because Evert and Ilma met while working in the Bloomer School System. Ilma Zemper was principal of a Bloomer elementary school when Evert arrived to become the high school vocational agricultural instructor and Future Farmers of America enthusiast.

The Bloomer experience was a real conversion for Wallenfeldt because he was from Iowa and a graduate of Iowa State College, of course. While Iowa had its hogs, Wisconsin had a dairy emphasis.

His experience in Wisconsin helped convert Evert to a nationally recognized person in dairying. And his abilities grew when he and his young bride left Bloomer so Evert could study for his master's degree in dairy science at Cornell University in Ithaca, N.Y.

Ilma was a boost all the way because she took typing courses so she could type Evert's thesis, which attracted the interest of the Borden Company. So for eight and a half years, the Wallenfeldts found their niche in the dairy business.

Borden quickly recognized Evert's talents and moved him from his first job as a fieldman in Illinois to Syracuse, N.Y., where he became a bacteriologist in Borden's main research facility.

Through his work, Evert was eventually attracted

to UW-Madison and he joined the faculty in 1938. His Wisconsin service was monumental and he was honored for his work in getting quality milk for a growing dairy industry.

Milo Swanton, a former secretary of the old Wisconsin Council of Agricultural Cooperatives, a federation of cooperatives based in Madison, once recognized Evert's work and what it meant for the state. The great mark of Evert, Swanton said at the time, "is that he showed personal willingness to cooperate and always give credit to others for their achievements."

Swanton said Evert was one of the few who worked untiringly for quality improvement, starting at the farm and dairy plant and motivating farm youth. "The dairy industry of Wisconsin and America owes much to Wallenfeldt, a devoted and highly effective servant of the dairy industry," Swanton said, in words that still ring true today.

Poem Celebrates Farm Christmas

December 25, 1994

It's Christmas and this is Professor Dave Dickson's season.

He is an extension dairy specialist at UW-Madison's College of Agricultural and Life Sciences and for 27 years has earned a new title — poet laureate of dairying.

Dickson credits his grandmother, who read poetry to him as a youth on a dairy farm in Washington state, as being a major source of literary inspiration. Over the years, he said, he polished his skills through a Christmas program heard over radio station WHA. He started with prose, like a letter to a farmer from the cows on things they'd like to see for the New Year. But then, to brighten the program, Dickson started to express himself poetically each year on matters of farm life around Christmas.

The WHA dairy program is gone, but Dickson has never faltered from his Christmas poem.

Dave Dickson

He has had his works published throughout the U.S. and used in every major dairy country — from Australia and New Zealand to Canada and Great Britain.

At home, Dickson's poem is part of Orion Samuelson's television Christmas program on WGN that is linked to scores of stations throughout the nation.

The following is Dave Dickson's 1994 Christmas message:

Holiday Dreams

Raindrops on hilltops and warm April showers,
Trim country farmsteads embroidered with flowers,
Songbirds that warble their musical themes,
Dance through my head in these holiday dreams.
Pansies and daisies and soft, gentle breezes,
Planting a garden the way Mama pleases,
Plowing straight furrows with dapple gray teams,
Mem'ries of springtimes bring holiday dreams.
Clover-filled meadows and new puppies drowsing,
Fawn-colored Jerseys and Guernseys just browsing,
Strawberry toppings for home-... made ice creams,
Flavor the taste of these holiday dreams.
Corn stalks and hay shocks and thunderheads clapping,
Blue-ribbon Holsteins and kittens cat-napping,
Dog days of summer are endless, it seems,
Shimmering bright in our holiday dreams.
Oak trees and maples, their reds and golds blazing,
Brown Swiss in pastures, contentedly grazing,

Still picking corn while the harvest moon gleams,
Thankful and proud of our holiday dreams.
Cider and pumpkins and frosty Novembers,
Plump roasted turkeys that last 'til Decembers,
Freshly cut firewood, hot chocolate that steams,
Warming our hearts and these holiday dreams.
Christmas surprises and sleigh bells that jingle,
Stockings filled up by a jolly Kris Kringle,
Crisp winter mornings and ice- crusted streams;
Pausing to ponder these holiday dreams.
Yule logs and carols, nostalgic reflections,
Old-fashioned New Years and fond recollections;
Brightly the splendor of hopefulness beams,
Wishing great joy for your holiday dreams.

Wisconsin's Poet of the Dairy Offers his Christmas Rhymes

December 23, 1990

The poet of the dairy farm, Prof. David Dickson, has been writing an annual Christmas poem since 1968 and his 23rd offering carries many messages.

Dickson, a UW-Madison Extension dairyman, sees Christmas as time for a gift to Wisconsin's 1.7 million milking cows and the hundreds of thousands of heifers that one day will stand in the milk lines of America's Dairyland.

He was raised on a dairy farm near the Canadian border in the northwest corner of Washington state and probably was in the fifth grade when he started to write in rhyme.

Dickson recalls how, as a youngster growing up in the pre-television age, one of his grandmothers could, from memory, recite poetry for hours — "really thousands of verses." He still has tapes of her reciting the poems the family loved to hear.

He caught the rhythm of iambic pentameter and many verses later he was certain that poetry was a lost art. Dickson says a poem can easily leave a subliminal message that lives in a person's mind for years. That's what he incorporates in this year's poem for those who care for cows.

There are some dairy terms that may need to be defined.

For instance, DHI, is Dairy Herd Improvement and its milk testing for production records; ET is embryo transfer, the placing of embryos from a high producing cow in brood cows to sustain and give birth to the high-producing calves; TMR is total mixed ration, a large drum for mixing and blending the necessary feeds and supplements for dairy cows and, by far, the most popular new item on today's dairy farms for efficient milk production; CFP, is combined fat and protein ratio used to measure the fat and protein transmitting abilities of sires; and cloning, which duplicates the genetic material to create two or more identical animals.

Here's Dickson's poem, and Merry Christmas and the New Year's best to you.

'Twas a week before Christmas and all 'round the farm,
the cattle were rustling in nervous alarm.
No stockings were hung, no halls decked with boughs,
just hopes that Saint Nicholas cared about cows.
The livestock were restless tied in the sheds,
while visions of thunderclouds danced in their heads.
Well, ma in her sweatshirt, and me in my jeans
had finished washing the milking machines.
When out of the barn came a puzzling clatter,
that left us both wond'ring what might be the matter.
I opened the door to the barn pretty fast,

and flipped on the lights as the clamoring passed.
The haze on the ceiling and whitewash below
gave each slumb'ring cow an ethereal glow.
Then, just as the dust and the murkiness cleared,
this little, old elf from the shadows appeared.
He was dressed in bib overalls, just like I wear,
with a snowy white beard, and long, flowing hair.
His pace was determined, his movements so quick,
I thought for an instant it might be Saint Nick.
He strode 'cross the barn, past the mangers he came,
and he called to each cow by her very own name,
"Up Wimpy! Up Gina! Up Roxy and May!
Come Charity, Ella! We haven't all day."
He opened each stanchion, then paused by a stall, crying,
"Dash away, dash away, dash away all!"
Then in a twinkling he unbarred the door,
and chased out the cows that had stood there before.
As I drew in my breath and was turning around,
he pointed his finger and spoke with a frown,
"You need some help quickly; these are tough times;
better start pinching those pennies and dimes.
"With milk prices tumbling, it's time you began.
Your cows are so worried, they've drawn up this plan."
He opened his ledger, them turned with a jerk,
and gave me a list that my cows thought would work:
"Let's try DHI, we're sure you'll agree,
there's profit in knowing to cull or ET.
Make alfalfa silage that's cut early but,
we'll give cleaner milk if you clear up the mud.
"Milk quality premiums will "beef" up your check,
and how about trying a little 'high-tech?'
That TMR mixer could save you some fuss;
please find us some sires that are CFP plus."
With cloning and markets available now,

the future's today for tomorrow's top cow.
Here's a book about inputs, expenses and such;
this list from us cows wouldn't cost very much."
Then, closing his notes with a flick of his thumb,
that gentleman vanished the same way he'd come.
It must have been Santa, this droll, little guy,
deliv'ring these changes the cows want to try.
Christmas means finding the things we can give,
then granting those gifts, this season that makes
life fun to live.
It's giving, not getting, this season's about.
I'll give what my cows have been doing without.
As I walked toward the house, sailing straight
into view, flew a miniature sleigh, and
Saint Nicholas, too.
And I hear him exclaim, as he whirled out of sight,
"remember your cows, make their holiday bright!"

Professor Peddles Poetry

December 26, 1993

It's Christmas and this is the place for the poet laureate of cows, UW-Madison dairy scientist Professor David Dickson, who now has written one yuletide poem a year for 26 years.

His words of rhyme give young and older Dickson fans a chance to think about cows and farms in a different light.

Now, it is proposed that Dickson's yuletide poems be published in a book so his readers can look back through the years and smile or laugh at the situations he describes.

Those interested in the idea are looking for some indication from his fans on whether a book of Dickson's poems, called "Christmas in Dairyland" should be published for the 1994 holiday season.

Dickson, who is from a dairy farm in Washington State, is not only a poet, but a respected researcher and head of the farm man`agement dairy program in the department of dairy science.

Dickson has been in Madison since 1960 and when he finished his master's and doctoral degrees, he was appointed a member of the dairy science faculty.

He has served as chairman of the department and is known throughout the state as an exceptional judge of dairy cattle, a special speaker, talented musician and artist with lively chalk talks at many public meetings.

But in any forum, he leaves a message that people are not about to forget.

And they do not forget Dickson.

So here, with wishes for your Merry Christmas and New Year happiness, is Dickson's latest poem — lesson 26 in his poetic series about cows and everyone who depends on them:

White Christmas (with black spots)

A frosted window glimmers with a twinkling candle glow,
While shadows bustle quickly, figures wand'ring to and fro.
Elves busy wrapping presents at a record setting clip;
'Tis Christmas Eve and Santa's set to make his annual trip!
The reindeer prance excitedly and shake their antlered heads
as Santa tests his radar, checks the toddlers in their beds.
He verifies his flight plan, feels a tug, then turns around;
Gently pulls, then yanks and hears a tearing, ripping sound!
"Good grief!' he says to Rudolph, his smile a bit

askance,

"Better hustle to the house so Mrs. Claus can mend my pants!

"When he dashes to the tiny house, he feels a drafty breeze

Come through the tear, now widened from his pocket to his knees.

And, by the time poor Santa comes careening through the door

His britches, far beyond repair, are dragging on the floor!

"It's lift-off time," he hollers, as he fumbles with the latch,

"I'll need my second pair of pants or else a fancy patch!"

"We haven't got the smaller pair, I threw them out because,

You've gained an inch or maybe two," reponded Mrs. Claus.

"But, never fear, I'll make new pants that fit you to a tee!

I found the perfect fabric with a pattern you should see!

"While shopping at the North Star Mall, I noticed just last week,

How all the stores are stocked this year with trendy Holstein chic!

Spotted white and black, I think, or maybe black and white,

From socks to mugs and tableware, a wondrous dappled sight!

Cotton throws and cookie jars, black mats, bowls and spoons,

And clocks with little Holstein cows high jumping over moons.

Oven mitts and chef's hats neatly spotted white and black;
Complete arrays of shirts and ties and shelves of bric-a-brac!
"I saw wind socks, even slippers, also pads and Parker pens,
A whistling little kettle and boxer shorts (just men's).
There were canisters and place mats, carafes and dominoes,
Just make it white with spots of black and sell it, I suppose!
"It's the fashion, its outrageous! Oh, so stylish, so genteel,
I had to purchase something with that dairy cow appeal!
So I bought a bolt of fabric, boldly spotted black and white —
I'll stitch a pair of trousers you'll be proud to wear tonight!
But, Santa, knowing well his reputation for respect,
Insisted, first, that cows should be politically correct.
"I've checked," said Mrs. Claus, "cows are wonderful, indeed!
Giving, working day and night for humankind in need.
"Producing butter, milk and cheese or pulling rustic plows;
Shoes and glues and sausages — all gifts from dairy cows.
They come in many colors — black or orange, red or brown,
Yet they live in peaceful harmony, wherever cows are found.
"Concerned for the environment, their culinary taste
Includes what many industries must leave as solid waste;

From bakers, canners, brewers — in her simple bovine way,
The cow consumes what humans can't, recycling every day.
"There you go, the pants are done: they'll make you proud as punch.
I've let the waist out just a bit, to leave some room for lunch!"
Santa steps into his trousers, festooned in dairy splendor;
A fashion plate, in Holstein pants held up by red suspenders!
His sleigh and reindeer rise above the trees, then briefly pause;
And a voice cries out from far below, a shout from Mrs. Claus,
"You look so dapper, Santa, in those spots of black and white,
Let Christmas 'round the world begin with dairy cows tonight!"

Chapter 9

In Memorium

Wisconsin Dairying Loses Friend, Leader in Death of Cooperative Giant

November 8, 1992

A decisive era of Wisconsin and regional dairying ended recently with the death of Melvin H. Sprecher, 81, of Sauk City.

Sprecher was a dairy farmer on a 255-acre farm near Witwen in Sauk County and with his fine country manner became a leader of dairy farmers when leadership was most needed.

He kept informed on all aspects of the business, but his greatness came with his unique ability to bring farmers together for their own good and that of the industry.

He accomplished that in the formation of Wisconsin Dairies Cooperative — the consolidation of the Wisconsin Cooperative Creamery Association at Union Center and the Wisconsin Creamery Co. Cooperative at Sauk City.

This gave dairy farmers an opportunity to realize the gains that could be made when cooperative dairy enterprises teamed up for the common good.

He was a leader farmers could trust.

Sprecher was able to provide a broad view of dairying, which stretched from the farm gate to the international arena, well before the move to the world marketplace became a positive force for farm decision makers.

His spirit guided the Wisconsin Dairy Federation and he served with distinction as a board member of the National Milk Producers Federation.

What Sprecher and his group of dairy leaders put together — in March of 1963 — was the sound and productive organization that became the base of Wisconsin

Dairies with its 2,000 members, which developed into one of the nation's top 500 business firms.

Della and Melvin Sprecher

Sprecher's talent for creative dairying and the cooperative food enterprise also was shown when he became president — and later chairman of the board — of Land O'Lakes Inc., when it was one of the 200 biggest firms in the nation.

In his talks to 2,400 Land O'Lakes delegates and to his Wisconsin Dairies members, he repeatedly maintained that the cooperative can give farmers the advantage of a total food production-processing and marketing operation that can help the family farm concept thrive.

All this built respect for the dairy farmer, who with his wife, Della, had to turn the cream separator by hand so the cream could be sold to Honey Creek Valley Dairy Cooperative at Witwen.

She was with Sprecher wherever he traveled, whether to the World Food Conference in Rome or the sessions around the Wisconsin Dairies area in Wisconsin or the numerous Land O'Lakes meetings in the Twin Cities.

He never shied away from the tough decisions dairy farmers had to make, like putting together a milk promotion and research program so every producer was a contributor and letting farmers know that the time had come when Land O'Lakes — the king butter marketing firm — had to include the manufacture and sale of oleomargarine.

When Sprecher retired as chairman of Land O'Lakes, he

was the dean of cooperatives, and was named by Congress as "Senior Statesman of Agriculture" in a presentation by Rep. Thomas Foley, D-Wash.

Sprecher, at the podium with Foley, looked lovingly at his wife and said if it all had to be done again he would marry the same woman, "because I believe that to be a successful farmer and cooperative director, you need the help of a good wife."

Farr is Remembered as 'A Friend of Dairy Farmers'

December 5, 1993

When Charles L. Farr, 73, died at his home in Madison Nov. 26, representatives of dairy farmers, their cooperatives and the proprietary dairy plants and many others connected to the dairy industry came to remember Charlie.

Retired dairy farmer and cooperative dairy and marketing leader, Harland Rue, Barneveld, spoke at Charlie's funeral as the representative of dairy farmers and gave Charlie the ultimate recognition as "a friend of dairy farmers."

Farr was an exceptional dairy economist. During his 21 years as director of the Dairy Division of the Wisconsin Federation of Cooperatives — 1964 to 1985 — he was the recognized public spokesman for dairying and sincere lobbyist for their cause.

This was a time when dairying went through a period of dramatically changing policies and programs. The structure of the industry also changed with an unprecedented number of mergers and consolidations in both the cooperative and proprietary segments of the dairy industry.

Charlie was a native of Nebraska and he got on the dairy track as a student in dairy manufacturing and agricultural economics. There wasn't a phase of the industry he didn't know in detail.

At one time he managed an ice cream plant in Omaha and later worked as an accountant with the Federal Milk Marketing Order in Omaha. He also worked in Sioux City, Iowa, where he met Ilene Nelson, who soon was to be his wife.

Charlie later developed marketing skills as an economist with Twin Cities Milk Producers Association, prior to moving to Madison.

The important role Charlie had in the dairy industry was his ability to explain to dairy farmers the intricacies of dairy policy. As Rue said, "We learned from Charlie — we agreed and disagreed with him."

Charlie also was an innovator of dairy programs, the most notable one he authored was the "standby pool."

Through this unique program, he was able to organize the reserve milk supplies from dairy plants in Wisconsin and Minnesota and then make that milk available to contributing dairies across the South and Southwest when their milk supplies where short, especially during the fall and early winter each year.

What Charlie's program did is bring Upper Midwest dairy cooperatives into a program that for the first time was able to allow participating cooperatives to tap into the class one (fluid milk) price paid in markets of the South and Southwest.

While the program was in operation it returned millions of dollars to Wisconsin dairy farmers whose cooperatives committed their reserve supplies of milk and made it available to cooperating Southern and Southwest dairies when they needed the supplimentary milk supplies.

Participants chose Charlie to run the program, which he was able to do in addition to his regular duties as Dairy Division director.

Charlie was in demand as a leader in major projects

because of his willingness and energy to take such important assignments and because of his ability to carry them out successfully.

None was so important as the time he was nominated by cooperative dairy organizations to coordinate a statewide information program for cooperative leaders and their patrons so they would understand the value of forming a new program built around support of each dairy farmer to carry out dairy promotion and reseach.

The key to his effort was letting the core of the dairy industry know what could be accomplished with a successful vote establishing a producer-financed promotion and marketing program for dairy products.

In a dedicated effort to get the message to dairy farmers and their leaders, Charlie successfully accomplished the mission that put Wisconsin in the lead nationally for the eventual National Dairy Promotion and Research Program.

When Charlie retired from the Dairy Division position in the Wisconsin Federation of Cooperatives he was named a consultant with the Wisconsin Milk Marketing Board to serve as a liaison with the farmers and dairy plants he knew so well.

Former Co-Op Chief Leaves Legacy of Service

March 31, 1996

MANITOWOC — I never thought I would be visiting Truman Torgerson so soon, but there I was on a Sunday afternoon in a long line winding out of a small funeral home where Torgy was at rest in a brown wooden casket.

I felt the reverence and respect of the men and women — young and old, from farm and city — who came to pay their respects to Torgerson.

He died suddenly at 80 years after an operation to give him new life through several heart bypass arteries. After a hopeful sign only a strong man could show that all had gone well, it didn't last. Torgy is no longer with us.

But he had prepared us for that sad moment.

He wrote two books on the life of Truman Torgerson, telling a story that started on his home farm in Dunn County. The tales recount his days at the Dunn County School of Agriculture from which he graduated. It was there he met Randall Swanson, school principal, who was a guiding force in Torgy's future through the UW-Madison and his future work in Extension.

Torgy's life told it all and really could be an example for coming generations living the good life of service. In 1939 he was a graduate of the University of Wisconsin College of Agriculture, and many of his UW boxing fans can still hear the cheers from the crowd of 15,000 in the Field House, where he won the 1939 light heavyweight NCAA crown by defeating Americo Wozciesjes of Syracuse.

But there also were cheers and admiration for his lifetime mission as president and general manager of a major dairy cooperative, Lake to Lake Dairy Cooperative at Manitowoc. It was an organization that worked for the good of the people who cooperatively owned the business he ran for 35 years.

It all began when Torgy was county agent in Manitowoc County, and three dairy farmers stopped in to see him with the hope that something could be done by working together to improve their lives.

First, Torgy turned to the University of Wisconsin College of Agriculture and agricultural economists Rudolph Froker and Marvin A. Schaars for help. But his plan wouldn't have happened without Torgerson's friend and adviser, Randall Swanson, then principal of the School of Agriculture, now of Madison and an emeritus UW professor.

How good was it all at Lake to Lake? The cheese became the first in the U.S. to qualify for the U.S. Department of Agriculture Grade AA shield as the highest quality cheese in the nation. This came through the efforts and contributions to excellence by Professor Walter Price, the UW cheese expert.

Even today, Lake to Lake Cheese is still identified on the cheese package label in stores throughout the nation, even though Lake to Lake was acquired by Land O' Lakes when Torgy retired.

Thus the name and reputation of a masterful dairy and cooperative leader lives on for the way he served his dairy producers, worked with dairy farm leaders and maintained his product quality.

He now rests with his family from the home farm and relatives in the town of Peru Cemetery in Dunn County.

Rock County Mourns Ziegler

May 13, 1990

There's been a void in Rock County's youth, dairy, and livestock circles since the recent death of Bob Ziegler.

Ziegler, 54, grew up on a dairy farm east of Madison and after graduation from Madison East High School and the UW-Madison he became one of the exceptional persons of Wisconsin Extension for his work with young people and the adults of livestock associations.

He died when stricken with a heart attack May 2.

Ziegler was an associate professor of Wisconsin Extension and was in his 20th year as dairy and livestock agent in the Rock County Extension Office.

Rock County residents, both youth and adults, respected Ziegler, calling him a leader everyone liked. He had informed everyone he was counting the days until his retirement next Jan. 3.

And shortly before his death and months before his planned retirement, Barb George, Brooklyn, said people were beginning to plan what kind of a retirement party and gift of remembrance Ziegler should have.

"The thing about Bob that tells why he was so well accepted by livestock people," she said, "is that being a livestock agent wasn't just a job for Bob — it was his life. His clients, young people and adults, were his best friends."

Marion Barlass, Janesville, said Ziegler's quality work will continue because he had so many good volunteer people working with him.

"He made his job as agent what he wanted it to be. Even though he was not coaching the dairy judging team, he came to their evening workouts to see if everything went right, that arrangments were made for the next weekly session and every team member knew how to get to the host farm," she said.

Ziegler came to Janesville in 1971 after serving as 4-H and youth agent in Sauk and Walworth counties.

He got his start in Extension as a 4-H summer agent in Dane County in 1957 and 1960. His studies in agriculture and animal husbandry at the UW-Madison were delayed by service in the Army for two years. He received his master's of science degree in 1971.

Ziegler gained friends and admirers in his work, Paul George said, because of his ability work with youth in livestock projects and on judging teams so they could learn something and profit from the experience.

He had a strong part in building the Rock County 4-H Fair's annual Livestock Sale. Together with the fair board and livestock leaders he developed and promoted the sale, which become the largest meat animal sale in the state.

Last summer 800 animals went through the sale rings, Paul George said, and returned $260,433 to the young exhibitors.

It wasn't the money that distinguished this sale, but the way people showed support for the county's livestock program as an incentive for youth. Ziegler also used the sale as an educational tool for youth exhibitors.

Paul George said Ziegler was working with county livestock groups to develop a plan to grade the sale animals, so all youth exhibitors would know the reason some steers, for instance, returned 70 cents a pound while others of different quality and grade did not command the price.

Ziegler coached the county 4-H livestock judging teams for beef, swine and sheep and his team members brought 11 state championships to the county.

Rock County Agent Dennis Nehring said Ziegler was his best friend and he found him "a most caring individual — a people-person who was very concerned about people and their lives."

Ziegler never married and Nehring said the people in the Extension Office and all those he served were really his family.

"He helped breed and livestock associations with their promotions. Night meetings and working weekends were things he really wouldn't have had to do, but he did it with zeal," Nehring added.

Co-op Pioneer Harris Left Mark on Dairy Industry

June 3, 1990

Friends in Wisconsin have learned of the death in Florida of Floyd Harris, retired first general manager of Wisconsin Dairies Cooperative in Baraboo and a major contributor to dairy technology.

He was manager of Wisconsin Dairies when he bought the French Simone, the first continuous butter churn in America, for the merged dairy operation.

The butter-making system Harris introduced is now used throughout America, churning 95 percent of all butter made in the nation.

A self-taught dairy engineer and butter expert, Harris worked closely with the late Prof. L.C. Thomsen of the UW-Madison dairy science department and dairy consultant Nick Fabricius.

With UW-Madison, Harris helped develop processes to make products such as sterile milk, ice cream, dry and liquid dairy blends and many types of cheese.

His quest for better butter-making systems resulted in the purchase of a German-made Ben Hil high-speed soft butter packaging machine in 1965 for the Union Center plant. This machine and improvements made over the next three years made it possible for Wisconsin Dairies to package 18,000 quarter-pounds of butter per hour. What Harris saw as a possibility in 1965 now is used to package about 75 percent of all butter in the nation.

He also was a leader in bringing three cooperatives into one building, the Cooperative Service Center, serving Wisconsin Dairies Cooperative, Equity Cooperative Livestock Sales Association and Tri-State Breeders Cooperative.

Harris conceived the idea as computers started moving

Floyd Harris

into the business world and the three organizations shared the same computer and office building. Wisconsin Dairies since has constructed its own larger office next to the center, north of Baraboo on Highway 12.

Harris, a former Baraboo resident, was born on a farm at Mount Tabor, near Hillsboro. A graduate of Hillsboro High School, he was 20 years old when he started in the dairy business at Hillpoint Cooperative Creamery.

He started as a milk tester, separator helper and butter maker. After seven years, he was promoted to manager of Hillpoint. He was the manager there for 15 years and was in charge when Hillpoint bought Nestle's plant in Reedsburg.

In 1950 he became manager of Wisconsin Creamery Cooperative in Sauk City. In that job he developed a system for the first liquid-petroleum-gas, direct-fired milk drier in the Midwest.

After the 1963 merger of Wisconsin Creamery Cooperative at Sauk City and Union Center's Wisconsin Creamery Association, Harris handled the administration of Wisconsin Dairies. During its early growth period, the dairies grew to an organization with 14 plants.

He died April 30 on his 83rd birthday in Titusville, Fla., where he and his wife, Bernadine, moved after his retirement. He suffered from cancer for a short time. He was buried in Loganville.

Melvin Sprecher, the first president of Wisconsin Dairies, said Harris was an excellent administrator "who could get his staff and employees to do things for the good of the cooperative.

"He was a dedicated cooperative leader who fully backed the formation of Wisconsin Dairies through merger because he believed farmers could benefit from such organizational strength," said Sprecher, of Sauk City.

Throughout his career as a cooperative manager, Harris had a knack for hiring capable staff. He also was able to work with large dairy companies, such as Kraft and Bordens, to supply them with the dairy products they needed.

Kirschbaum Served Agriculture, Consumers Well

May 19, 1996

The recent death of Norman E. Kirschbaum, who served both agriculture and consumers for 35 years in the Wisconsin Department of Agriculture, Trade and Consumer Protection, brought back memories of what his service meant to Wisconsin and the nation.

Groups that enjoyed his talent, counsel, and exceptional good nature, paused during their recent meetings to memorialize Norm's service.

It was a special moment of remembrance for an exceptional leader, who knew how to achieve reasonable solutions in most difficult circumstances.

These were assignments in animal health, during the opening of ring tests to eliminate brucellosis and later bovine tuberculosis in Wisconsin, when there still were 152,000 dairy herds in the state, compared to 28,000 at present.

This was well before electronic records and each animal's test was recorded by hand as the state worked through more than nearly 2 million cattle to eliminate the diseases.

As a result of his mastery of record keeping, organizational benefits, and his ability to work with dairy farmers in times of stress and some controversy over the loss of infected cattle or even entire herds, Norm worked his way to an appointment as administrator of the Food Division of the department.

He served that post for 18 years and retired in 1985, ready to serve as executive director of the Wisconsin Dairy Products Association for five years. He then became the agricultural specialist for Sen. Robert Kasten, and received an appointment by Gov. Tommy Thompson to the World Dairy Center Authority, dedicated to the expansion of the center's services for affiliated business organizations.

In his professional career, he had to settle state differences between farmers and food concerns. But former Deputy Ag Secretary Art Kurtz recalled that Norm always had a disarming smile as he dealt with the principals involved.

It was the communication of a smile, which put the two sides in position to settle major problems with a negotiated agreement, Kurtz said.

Norm's ability was recognized in national organizations, including a key election as national president of the Association of Food and Drug Officials.

His record as an administrator resulted in Norm being a recipient of the national "Harvey W. Wiley Award."

He also served his church, Madison's Our Lady Queen of Peace, and its organizations to further extend his life of service to men and women of town and country.

Lyman McKee's Influence Reached Beyond the Farm

June 13, 1993

Lyman D. McKee died on June 3 at 81 years, but his leadership and skills as a dairy farmer and dairy marketer nationally and internationally, will live on.

Those of us who, for decades, had a chance to watch Lyman at the podium and on the farm greatly admired him because he conducted himself in a way that benefited everyone who worked with him. They often wished that more of Lyman's charm would have rubbed off on them.

He had the magic of Irish humor and an engaging chuckle that made you laugh with him at his own jokes.

Lyman could come up with good ideas, especially the one creating the first Dairy Farmers Retirement Plan by having members contribute regularly to the fund through their milk checks. He was a master of fund investment and one year increased the fund with a boost of a 20 percent interest rate. The farmer recipients live happier lives today.

Dairy farmers now are having a hard time with health insurance payments and his dairy associates still say that if this had happened 25 years ago Lyman would have developed a workable health program for them.

But most of all, Lyman always emphasized service to dairy farmers, their markets, customers and the dairy industry. Those were the esssentials of the cause he worked to achieve.

Consequently, he never pushed himself to the forefront. He didn't have to because he was there already and everyone knew it and respected him for it. This was demonstrated as he was recognized with nearly every award bestowed on the greats of agriculture, dairying and rural leadership.

People always were amazed with the way he was able

to move into cooperative dairy organizational work through Madison Milk Producers Cooperative.

His father, Edward J. McKee, also a dairy farmer, was a director of the cooperative. After finishing a tour of Navy service, Lyman and his brother, Bill, took over the home dairy operation and farms as McKee Brothers. He had a keen interest in milk processing and marketing. To gain more information, he always went to the directors' meetings with his father.

When his father wanted to retire from the board, who was there to take his place?

Lyman.

That was not a matter of selecting Lyman because of his father. It was because the board knew that Lyman was as knowledgeable on issues facing the cooperative as any board member. He had gained such insight just by sitting through all those meetings.

It wasn't long before Lyman, from Route 3, Madison, in the area that now is part of the city of Fitchburg and where the development was on McKee farmland, was elected president of the cooperative. Later, his skills were such that he held managerial posts with Associated Milk Producers Inc. after Madison Milk merged with AMPI.

Through the cooperative, he attracted statewide attention and Lyman was elected president of the American Dairy Association of Wisconsin, which used voluntary contributions of dairy farmers to begin dairy promotion programs. He served as state ADA president for 12 years and then was national ADA president.

At the right moment, as milk production increased and domestic markets could no longer handle the growing supply of dairy products, Lyman became the first advocate and organizer of international marketing of dairy products.

He was innovative in such a move. He worked with the Wisconsin Farm Bureau Federation and commodity specialist Frank Wing to team up with dairy farmers, cooperative and proprietary dairy business people — those who could seize the initiative to make a dairy product sale when the opportunity presented itself. They made a world tour to seek markets for dairy farmers at home.

He never gave up the challenge of gaining international dairy markets and for years worked intensely with Dairy Society International.

The world began to see Lyman's ability in action on trade missions and trade talks. He also gained the attention of foreign leaders when he served as a delegate to a United Nations food conference in Beirut.

Members of the British Milk Marketing Board came to visit Lyman at his farm after meeting him in England. For a year and a half, Lyman lived at St. Mary's Care Center and had a typewriter handy, giving him a chance to write to his friends and recall some memorable moments of dairying that he helped create.

He also wrote a history of Fitchburg in classic McKee style — accurate, significant and memorable.

Index

A

B

C

F

G

H

I

J

K

L

M

N